T0250380

The Web and Faith
Theological Analysis of
Cyberspace Technologies

RIVER PUBLISHERS SERIES IN INFORMATION SCIENCE AND TECHNOLOGY

Series Editors

K. C. CHEN
National Taiwan University
Taipei, Taiwan
and
University of South Florida, USA

SANDEEP SHUKLA
Virginia Tech, USA
and
Indian Institute of Technology Kanpur, India

Indexing: All books published in this series are submitted to the Web of Science Book Citation Index (BkCI), to CrossRef and to Google Scholar.

The "River Publishers Series in Information Science and Technology" covers research which ushers the 21st Century into an Internet and multimedia era. Multimedia means the theory and application of filtering, coding, estimating, analyzing, detecting and recognizing, synthesizing, classifying, recording, and reproducing signals by digital and/or analog devices or techniques, while the scope of "signal" includes audio, video, speech, image, musical, multimedia, data/content, geophysical, sonar/radar, bio/medical, sensation, etc. Networking suggests transportation of such multimedia contents among nodes in communication and/or computer networks, to facilitate the ultimate Internet.

Theory, technologies, protocols and standards, applications/services, practice and implementation of wired/wireless networking are all within the scope of this series. Based on network and communication science, we further extend the scope for 21st Century life through the knowledge in robotics, machine learning, embedded systems, cognitive science, pattern recognition, quantum/biological/molecular computation and information processing, biology, ecology, social science and economics, user behaviors and interface, and applications to health and society advance.

Books published in the series include research monographs, edited volumes, handbooks and textbooks. The books provide professionals, researchers, educators, and advanced students in the field with an invaluable insight into the latest research and developments.

Topics covered in the series include, but are by no means restricted to the following:

- Communication/Computer Networking Technologies and Applications
- Queuing Theory
- Optimization
- Operation Research
- Stochastic Processes
- Information Theory
- Multimedia/Speech/Video Processing
- Computation and Information Processing
- Machine Intelligence
- Cognitive Science and Brian Science
- Embedded Systems
- Computer Architectures
- Reconfigurable Computing
- Cyber Security

For a list of other books in this series, visit www.riverpublishers.com

The Web and Faith
Theological Analysis of Cyberspace Technologies

Ayse Kok

Google
USA

River Publishers

Published, sold and distributed by:
River Publishers
Alsbjergvej 10
9260 Gistrup
Denmark

River Publishers
Lange Geer 44
2611 PW Delft
The Netherlands

Tel.: +45369953197
www.riverpublishers.com

ISBN: 978-87-93609-57-0 (Hardback)
     978-87-93609-56-3 (Ebook)

©The Editor(s) (if applicable) and The Author(s) 2018. This book is published open access.

Open Access

This book is distributed under the terms of the Creative Commons Attribution-Non-Commercial 4.0 International License, CC-BY-NC 4.0) (http://creativecommons.org/licenses/by/4.0/), which permits use, duplication, adaptation, distribution and reproduction in any medium or format, as long as you give appropriate credit to the original author(s) and the source, a link is provided to the Creative Commons license and any changes made are indicated. The images or other third party material in this book are included in the work's Creative Commons license, unless indicated otherwise in the credit line; if such material is not included in the work's Creative Commons license and the respective action is not permitted by statutory regulation, users will need to obtain permission from the license holder to duplicate, adapt, or reproduce the material.

The use of general descriptive names, registered names, trademarks, service marks, etc. in this publication does not imply, even in the absence of a specific statement, that such names are exempt from the relevant protective laws and regulations and therefore free for general use.

The publisher, the authors and the editors are safe to assume that the advice and information in this book are believed to be true and accurate at the date of publication. Neither the publisher nor the authors or the editors give a warranty, express or implied, with respect to the material contained herein or for any errors or omissions that may have been made.

Printed on acid-free paper.

Contents

Preface   vii

List of Tables   ix

List of Abbreviations   xi

1   Defining Technology   **1**
     1.1   Introduction   1
         1.1.1   Technology or Soulcraft?   3

2   Physical Dimension of Technology   **7**
     2.1   Introduction   7
         2.1.1   Thinking Machines   13

3   Spiritual Dimension of Technology   **23**

4   Communicational Dimension of Technology   **43**

5   Occupational Dimension of Technology   **49**

6   Trade Dimension of Technology   **65**

7   Dark Dimension of Technology   **71**
     7.1   Introduction   71
     7.2   Encryption Runs Counter to the Business Interests
         of Many Companies   76
     7.3   Other Surveillance Mechanisms: The Internet of Things
         and Networked Sensors   77

8   Final Remarks   **81**

9   Appendix A: Detailed Overview of Innovation and Competition Policies   **83**
9.1   Introduction   83
9.2   Taxonomies of Innovations   83
9.3   Literature Review   85
    9.3.1   Theories Underpinning Competition Policy   85
    9.3.2   Theories Underpinning Innovation Policy   86
    9.3.3   The Role of Competition Policy in the Innovation Process   87
    9.3.4   Competition in the Market or for the Market?   88
9.4   Theoretical Results   90
    9.4.1   Competition versus Monopoly   90
    9.4.2   Racing Models   90
    9.4.3   R&D Cooperation Policy   91
    9.4.4   R&D Subsidies   92
    9.4.5   Patents   93
9.5   Empirical Evidence   94
    9.5.1   The Rate of Return to Investment in R&D   94
    9.5.2   Cross-sectional Evidence   95
9.6   Suggestions   96
9.7   Conclusion   97

10 Appendix B: Conceptualizing Cyber Security from the EU Perspective   **99**
10.1   Concepts and Approaches of the EU's Cyber Security Policy   100
10.2   The EU Approach to Cyber Security   102
10.3   Progress and Challenges of the EU Cyber Security Strategy   103
10.4   Conclusion   105

11 Appendix C: International Code of Conduct for Information Security (SCO)   **107**

References   **109**

Index   **115**

About the Author   **117**

Preface

No one would deny the fact that when one claims to be living a modern life, digital technology would presumably have become an indispensable part of this modernity. Given the fast-paced changes within this modern way of living, no one would deny the remoteness of this modern way of living from the way of living narrated in the scriptures. The rapid proliferation of information communication technologies into our daily lives has also affected the dynamism of the religion. Apart from the question of when and whether growth takes place for the right reasons, it seems to be the case that those institutes – including religious ones – that embrace technology in an innovate way may spread in the fastest way in the so-called global religious marketplace. As conferences are simulcast, scriptures are promoted on blogs, prayers are live-streamed, and religious conferences are simulcast, religious scholars also get exposed to millions of their followers on social media in a digitally saturated society.

On the other hand, this pervasiveness of computer technology does not mean that human beings have found a way to live as faithful persons in our digital age. Being aware of this lack of awareness, I intend to provide a faith-based perspective of technology rooted mainly in the Islamic tradition. Needless to say, technology is neither neutral nor provides a "sacred" means of digital use. Therefore, I intend to guide the reader of this book to see the digital world as part of God's creation and to use it responsibly based on the laws of God in order to make technology a useful and meaningful part of mankind's "struggle" ('jihad) for both this life and the Afterlife.

According to conventional wisdom, science and religion may stand in opposition to each other as there may be a mutual oppression between them in our modern times. Nowadays, most believers and scientists are too cautious of reducing religion to science or vice versa; we should rightly not trust those who seek only scientific evidence to prove religious truths or read the verses as textbooks of science. Yet, given the fact that the rise of science also depended on some religious virtues and metaphysical beliefs rooted in religion, this relationship is not simply antagonistic. While scientific inquiry has often been enriched by religious institutions, scientific insights into the natural world have also supported religious beliefs. Likewise, technological innovation has played a crucial role in the spread of religious ideas.

Likewise, conventional wisdom may hold that technology like science stands in opposition to religion. Sadly, the same technology that can empower individuals can also annihilate our spiritual senses or reduce our efforts to concentrate on prayer and meditation given its so many distractions. Despite the increasing level of technology in our daily lives and their negative effects upon our lives, individuals will continue to seek their Creator. As technological innovations can be utilized for serving noble or ignoble ends, technology professionals, like all other individuals, can utilize their skills for serving selfish or unselfish ends. Taking a faith-based perspective at technology, one will realize that there are abundant resources not only in terms of criticizing technology, but also in terms of its positive appreciation. With regard to this aspect of the faith, I will refer to the following two categories:

- How can we find our Creator in the work of technology or the aims that these technologies serve, and
- How can we find our Creator in technologies (See Appendix A for a complete review of innovation and competency policies) that serve humankind.

Writing from a faith-based perspective, this book touches on philosophy and history, and explores how technology and modern science indeed influence faith and religious practice. By drawing from a wide range of religious scholars and penetrating a cultural and theological analysis based upon the work of leading media scholars in this field such as Marshall McLuhan, Jacques Ellul, and Neil Postman, this book is a humble attempt to situate computer technology within the big picture of faith.

List of Tables

Table 7.1   Terrorist uses of the net   73
Table A9.1   Modes of innovation and possible policy responses
                 based on type of market failure   84
Table A10.1 Instruments of power in cyberspace   100
Table A10.2 Strands of EU's cyber security policy   102

List of Abbreviations

AI	Artificial Intelligence
CIP	Critical Infrastructure Protection
EEAS	European External Action Service
ENISA	European Network and Information Security Agency
FTC	Federal Trade Commission
ICT	Information Communication Technologies
IoT	Internet of Things
IT	Information Technologies
NIS	Network and Information Security
R&D	Research & Development
SCO	Shanghai Cooperation Organization

1

Defining Technology

1.1 Introduction

A technology refers to a technique or tool that enhances the individual power to manipulate one's environment. The word "technology" itself is derived from the Greek word of "tekhnologia."

Technologies are often conceptualized as referring to "high-tech" devices thanks to the developments within the scientific field. It may come as a surprise to most individuals that even the laptops in our backpacks may be considered as a kind of technology since it was developed by Dick Kelty in the1950s. Similarly, the pencil can be considered as a communications technology that was developed several hundreds of years ago and refined to its final form by a man serving in Napoleon's army in the 1790s.

According to some religious traditions, our Creator himself can be considered as the first technologist, given the clothing fashioned by Him for Adam and Eve on their departure from the Garden as a kind of "garments of skin." Technology's ambivalent nature is already reflected in this earliest technology in the form of being an artificial tool or substitute that functions as a supply to that which was lacking in our natural state. Moreover, Prophet Noah can be considered as the first carpenter of this world since he used to design his ark which not only acts as an unnatural covering made necessary by death but also serves as a sacred vessel for shielding others in their vulnerability. A range of technai – skills or crafts – include smithing, woodworking, and weaving as items are utilized for a meaningful design. Therefore, "techne" emphasizes the act of development as the individual is not completely expressing his art unless he is creating actual artifacts which enhance the lives of others by being useful.

Given various definitions of technology, the one co-developed by Challies and Dyer seems to be the most meaningful and relevant one given the scope of this book. These scholars refer to the meaningful act of tool development to transform God's creation for practical purposes as "technology." Given the fact that individuals have used tools to contribute to the flourishing of human beings since Adam, it is time to consider how Muslims might see technology as crucial to their deeds. What may be their main concerns as well as unique opportunities for service given their faith?

Some of us like me optimistic about our times perhaps because of the fact that we have not only been wired by God for possessing such a nature, but perhaps more due to truly believing in the sovereignty of our Creator. Given our unshakeable belief in God's sovereignty, we must live more responsibly and thoughtfully as we also believe in the consequences of our actions in the next world. In order to create a positive impact upon our times and culture in history in which our Creator has created us, we all must be familiar with the productive use of these technologies and tools of our times. There is no choice for us of opting out if as Muslims we truly believe that one of our major struggles is to impact the world in a positive way. As the Qur'an states, the proper foundation for our lives is the teachings of the Prophet Mohammad (s.a.w.w):

> *"Allah showed great kindness to the believers when He sent a Messenger to them from among themselves to recite His Signs to them, purify them and teach them the Book and Wisdom, even though before that they were clearly misguided." [3:164]*

> *"[The believers are] those who follow the Messenger, the unlettered Prophet, whom they find written down with them in the Torah and the Gospel, commanding them to do right and forbidding them to do wrong, making good things lawful for them and bad things forbidden for them, relieving them of their heavy loads and the chains that were around them. Those who believe in him, honor and help him, and follow the Light that has been sent down with him are successful." [7:157]*

At the beginning of history, as ancient communities were organized around evolving agricultural and communication technologies, the delicate combination of these technologies along with the physical labor constituted a crucial aspect of the monastic life. Being one of the 12th-century Augustinian scholars, Hugh of St. Victor shared his insights into the basic study of God in his work referred to as the "Didascalicon." Perhaps, technological improvement of the world may be part of our struggle in this world and the created order.

For ancient scholars such as Hugh and his followers, knowledge can be classified as follows:

- The practical (politics, economics, ethics),
- The theoretical (theology, mathematics), and
- The mechanical or technical (techne as medicine and fabric-making).

Classified as a fourth category, the field of logic, underpins all other three fields by providing certain rules for argumentation and rhetoric. According to these ancient scholars such as Hugh and Augustine, all knowledge leads to the Creator, so the technological quest shall also inquire into the natural order the Creator ordained, and thus into the Creator himself. In other words, technology has great potential for being spiritual as well. Despite the fact that most of the classical authors demeaned mechanical crafts, Hugh could harmonize the scriptural and technological understanding by treating technology as a means to compensate for our physical weakness. Our ability to serve the Creator and its creation is further enhanced by technology which allows humankind to become His servant in the natural world.

Being one of Hugh's followers, Godfrey asserted that the cultivation of mental disciplines is required for the mastery of the mechanical arts in order to be able to manage our thoughts and passions. This may seem contrary to one of the common criticisms of the modern day technology given the fact that its enjoyment often ends in distraction. According to Godfrey, the correct employment of technologies can help the mind to obtain the necessary focus and observation skills which influence the quality of good prayer.

According to the theologian Bonaventure, who used to live in the 13th century, the mechanical arts can be practiced to convey the grace of the Creator to others. To give a more specific example, the architect can be considered as providing a vessel of grace for the family who is going to live in the home he designed. So the mechanical arts can be conceptualized as instruments of both divine love and neighborly love. Early technologies such as quill and ink, paper and parchment – all have been important for the spread of religion.

Fast forward to the 21st century, we now have less ink and paper and more digital applications and video games. Nevertheless, they can still be utilized by taking into account the basic principles of faith.

1.1.1 Technology or Soulcraft?

The video game developer Blizzard Entertainment, best known today for its massively popular World of Warcraft (2004), first released a less known classic in 1998: StarCraft. The science fiction warfare and strategy game was the bestselling PC game of the year, and it sold nearly 10 million copies over the next decade. Professional competitions drew crowds of over 100,000 people in South Korea, where the game was so popular that three separate television stations regularly broadcast matches. Blizzard released a sequel, StarCraft 2: Wings of Liberty, in 2010 and a remake of the original, StarCraft: Remastered, in August 2017. From the right perspective, the checkered story of its creation offers lessons that extend far beyond the sphere of computer games and coding into our social and spiritual lives.

Coming off the heels of its success with the first two Warcraft games (precursors to World of Warcraft) in 1994 and 1995, Blizzard wanted to continue pumping out the hits every year. For those not old enough to remember, keep in mind that this was the time of CD-ROMs and dial-up Internet. The online world we take for granted and carry with us in our pockets today was in its infancy. Producing software far below our contemporary standards often took just as much, if not much more, work. Blizzard had high ambitions, to put it lightly. Patrick Wyatt, one of the lead game developers for StarCraft, told the story at length in 2012 on his blog Code of Honor and explained how with a limited timeframe and human resources, the StarCraft team's goal of developing a modest game could be described as "Orcs in space."

Another project (Diablo) from a newly acquired company (Condor, renamed Blizzard North) sucked away resources from StarCraft. As this Diablo project was extended in its scope all type of employees ranging from programmers to sound engineers working at Blizzard HQ dedicated their efforts to the game until StarCraft had no one left working on the project. Wyatt describes how the growing market for real-time strategy (RTS) games like StarCraft meant greater competition and how that competition pushed them to produce a better product. The unrealistic production schedule led to a cascade of problems. Personnel suffered from lack of experience at all levels, from junior coders to project leaders. StarCraft ran the risk of succumbing to what the economist Kenneth Boulding called the "sacrifice trap," where people continue to support a failed cause or relationship because they are too committed to its success.

One need not know anything about coding to share in the lessons Wyatt and his team learned along the way. Here, we see again a lesson that reaches far beyond the world of computer programming. Wyatt notes how the "many months of suffering" led him to improve his craft of coding for the better. So we may also note the ascetic benefit of enduring trials. Regardless of the level of sophistication of a particular tool or technology that is being developed – whether it is parchment or a video game – basic principles of faith are always in use. As stated in the Qur'an:

> *"Oh you who believe! Seek help with patient perseverance and prayer, for God is with those who patiently persevere." [2:153]*

There is something more here, however. As Wyatt implies, what the team needed to honestly stare their problem in the face, set aside the urgency of their self-imposed deadline, and build on a more stable foundation. As it has been stated in the Qur'an:

> *"Indeed, mankind was created anxious." [70:19]*

Taking the time to calm one's anxieties before acting leads to wiser and more effective (not to mention virtuous) actions. The Islamic tradition is abundant with resources for inspiration on how to enrich our societies in a more humane and developed way. The question that needs to be asked is whether we are inclined to recover those sources rather than being busy with enjoying the fruits of technology. Despite all the setbacks and missteps, StarCraft still managed to build upon a solid gaming foundation. Viewed from the lens of a spirituality of everyday life, even something ordinarily thought to be a great distance from religion, economics, and philosophy – computer games and coding – may contain gems of social and spiritual wisdom, if only we have eyes to see them.

2

Physical Dimension of Technology

2.1 Introduction

The world seems to be the provisioner of "content" for our social media reports. Without the individuals being aware of it, Facebook and Twitter have made their users view the world as containing lots of raw material to be delivered through means of various Facebook status updates and tweets.

While the Facebook prompt asks an individual "What's in your mind?" the Twitter prompt asks you "What's happening?" Needless to say, these two are subtly different. While Twitter asks you to act as a kind of news reporter of various sorts Facebook asks you to report on your inner mental states (What are you angry about? What are you sad about? What's in your mind this morning?). Yet, Twitter and Facebook prompts are identical if you consider that what is "happening" could very well be mental events rather than what is "happening outside." Between "what's in your mind"—what is inside there?—and "what's happening"—what's out there?—we seem to have all bases covered. There is no explicit collaboration between Facebook and Twitter, yet if these firms were partners then these prompts would collectively ensure a reasonably comprehensive elicitation of mental states and empirical reports.

Facebook users can "love" a post; they can be "angry" at what it talks about; they can be astonished and say "wow"; they can be "sad" and make a face dripping with tears; they can find it funny and laugh out loud "ha-ha." Now, a user does not just "like" a post; he can "react" to it. These, and others, are the templates that Facebook provides users for their reporting "work"; these are the ontological categories with which users are sent forth into the world to report back on what they see and hear and taste and feel. They make the world for users; they bid users look at the world with their lenses.

Those who are aware of the allure of Facebook "likes" would describe this pseudo-pleasure as hollow as they are seductive. The term "attention

7

economy," which refers to the evolution of the Web grounded on the demands of the advertising economy, has become a buzzword among Silicon Valley entrepreneurs.

Often times, despite the best intentions of individuals to develop tools, these may have negative consequences which were unintended at the beginning. In a similar vein, technology may result in making the users become addicted and pay only partial attention by constraining individual ability to focus to a great extent, which may eventually result in a lower IQ. According to one recent study, cognitive capacity may severely be damaged by the mere presence of smartphones even when the device is turned off.

When Facebook developed a path of least resistance by means of a single click in order to make some bytes of positivity flow across its platform, its "like" feature has been very successful. While Facebook users were enjoying the short-term boost they received due to their social affirmation, Facebook itself gathered valuable data with regard to its users' preferences which would become a great product to be sold to the advertisers. Soon, the same idea would be implemented by Twitter, with a slight difference of the heart-shaped "likes" and other websites and apps such as Instagram.

Facebook is constructing a map of the collective mind: at what time during the day did a user enter which status in response to which prompt; what online or offline event did it follow; what succeeded it; where was the user when he did so; what was he doing. Users are prompted to assist in the construction of this map, and they are complying. They are being interrogated; and they are complying. They are conditioned to do so; their responses are reflexive. In realistic art, they sought to capture the world as it was. In Facebook and Twitter art, individuals capture the world as expressed in status report and tweet.

These forms of social media work have granted users an incentive scheme of sorts. They can pour out the most "inconsequential" thought that flits across their mind and see whether their friends will "like" it or "react" to it or "pass it on" with stamped approval by retreating it.

Through this scheme, individuals have been provided a means by which they may seek confirmation whether there is any "value" to the constant stream of observations – all those thoughts and sensations – that pass through their minds. Also, this never-ending stream of data enables a great deal more than just friendly interactions among friends and family; every photo uploaded to Facebook and Instagram is harvested by bots and finds its way to the data banks of the NSA and the FBI and the CIA, there to be processed and used as learning data for face recognition software used to track criminals, or terrorists.

As individuals are being "asked," by social media, to make their interiority external, they seem to be happy to do so. There was a time when individuals needed to write the novel, with its richly imagined interior mental spaces, to enable this kind of inspection; then came psychoanalysis with its couch and undirected free association. Now Facebook's status feed and user timeline offers access to those reports, all made available voluntarily. Individuals log in, lie on their virtual couches, and chatter away.

Of course, while individuals rush to social media to tell the world what they saw, they often only tell the world what they saw if they think they will get enough positive feedback from it. Social media users see the activity on their friends' pages, how they attract admirers and compliments, and they crave the same. Everyone wants applause, cheers, and acknowledgments, and for that, they need the right "material," which the world and their friends provide. Social media users in general are popularity seekers, of the kind provided by their "friends" and "followers." There is a hierarchy of likes: the positive comment is still better than the like, and even better than the "love" emoji; the re-tweet is infinitely preferable to the "like." Users still prefer explicit communication to the terseness of the icon. By witnessing their friends' reactions to their own posts they realize which sentiments are safe to express, and which ones are not. They mold themselves according to the demands of their social media network.

Social media users seem to pay closer attention to the world as status and tweet composers; what their "friends" show them may reveal aspects of the world they had not noticed before. A map of the emotions, of the various affects and experiences, of the various changes in psychological dispositions during the day, is drawn for them; they are invited to aid in its preparation, to inspect it and comment on it. They feel compelled to provide their own maps in exchange. Yet, maps alter the world individuals see; they make them see a world in a very particular way; they are infected with selection biases all of their own. If reality is socially constructed, the Facebook status and the tweet are its new dimensions, its new axes of interaction and action. They offer users a perspective and a lens through which to view this world; they tell them what is to be condemned – that which gets the most "angry" reactions; what is funny; what is to be approved, which gets the most "shares," the most "likes," the most "RTs."

Social media users who report on their feelings or their observations are not merely hankering for approval; those two spaces now suggest themselves as precisely the places where such reports should be made. That is not just because they often do so; it is because they see others do the same thing, all the time, everywhere. A social media status is an act of social participation, an interaction with the world. If Socrates was right that "the unexamined life is not worth living," human beings seem to be in an era where the unpublicized act is not worth doing; or worse, did not happen at all. If something happens in the world and no one responds to it on Facebook or Twitter, did it really happen?

Now, individuals see the world differently, made up of check-in locations, of situations which need responses to. When they look at the world, they mostly see it through a social media filter: Is there anything here they could draw upon and use? What in this situation is amenable to formulation as Facebook status or tweet? They are overcome by the urge to report, to translate this "reality" into status or tweet. "Reality" is at its most disappointing when it does not present material suitable for usage in a status or tweet.

When an individual writes a Facebook status about an event – as opposed to writing a blog post about it – he shrinks the event into manageable form; he reduces its complexity, its many facets, and dimensions. Some users may write posts that are hundreds of words long on Facebook; the majority write short reports and one-liners. Many Twitter poets spend considerable time polishing their 140-character (which has just been extended to 280-character at the time of writing this book) tweets into a distillable aphorism. The picture of the world that emerges is of one that is capable of being captured so. The world is now the world as witnessed in social media, described and annotated in a very particular way, fitted into particular formats – those made available by the social media tools being used.

Within the realm of social media, most part of the communication, which may have been classified as being private in the past, has now become available to public spaces due to the feature of "sharing" among one's friends in various social media channels. There is a delicate line between privacy and social isolation which may run counter to our human nature. Real happiness occurs when there is conformity to the nature of human beings; on the other hand, this idea of conformity may sound provocative given the individual autonomy to a radical extent.

According to the social media scholar Peggy, the "exhibitionist culture" emerges as privacy keeps serving this isolation of the individual. Given this culture, we continue to get informed on things about each other that we have no right to know or that we should not know. Although technology may contribute to this phenomenon, the loss of personal privacy may also play a role in this. Due to this loss of privacy, the distinctiveness of the souls of human beings is also lost, which may eventually result in the deterioration of the civil society as well as its institutions.

Of course, this shall not mean that all privacy is bad. Without privacy, we cannot have our required solitude at certain times. For instance, solitude provides the individual with the opportunity to withdraw from the chaos of everyday life into a quiet place for prayer. Such moments enable one to re-evaluate oneself and to establish a balance for meeting the personal and professional demands of life. Within this regard, solitude contributes to a meaningful involvement in the broader society.

Given the current architecture of social networking technologies, behavior which may violate traditional privacy norms may be induced upon social media users. To give a specific example, Facebook's default settings allow information exposure to a maximum degree and for changing them, the user needs to go through a complicated opt-out process. This has been indeed the intention of its designers as new social norms are facilitated to share information so that in the end the modern networker becomes not much concerned about privacy issues as previous generations did.

Facebook's Wall would be a good example for its privacy-damaging architecture. Since the launch of Facebook, its Wall existed and users could become aware of how its privacy norms were revised. From its inception onward, Facebook encouraged its users to write on each other's wall which was quite different from sending a message. A message on the Wall acts as a kind of public message as the content can be seen by everyone. In a short period of time, a pattern emerged among the Wall messages as what was once written in email messages including planning dates, discussion of medical test results, etc., was now being written on Walls. In a similar vein, the comments spaces of Facebook posts provide an opportunity for communication which might have been restricted previously to email messages. Despite some revisions in Facebook's policies with regard to the user concerns on its security architecture, Facebook's default settings for information-sharing keep changing our collective understanding of privacy in social spaces (See Appendix B for a detailed discussion on this topic).

Needless to say, Facebook could easily defend itself against the charges of user privacy violations as the users themselves chose to walk into the trap set by Facebook.

In the past, the Internet has been conceptualized as a means for decentralized ownership and control. In comparison to the possession of natural resources, the ownership of technology and skills may be even more important when it comes to the establishment of knowledge economies.

Given the link between intellectual property and productivity, today's means of production have become knowledge, skills, and a capacity to fulfill the needs of others. One of the cornerstones of modern enterprise is the intellectual property as it is a crucial aspect of human creativity that may eventually result in economic prosperity to the benefit of all.

As the property rights are vital qua human rights on the national and international scale, the right to privacy is also crucial. To give a specific example, let's think how Google keeps all of its user data to more than 50 million users per day. Users "google" various things, get involved in online discussions, send emails, and make use of a host of various other services. Given this vast amount of information, how much of our own privacy can we forego for the sake of receiving better services?

There is something undignified and sordid about the whole business: this massive machinery of communication, with its complex software and hardware and intangible protocols, is dedicated to selling users goods. Individuals rarely remember that their communications with family and friends, their passionate and informed political discussions, are merely there to inform advertisers of user preferences. Something important happens on social media – users are, after all, communicating with each other – but they realize too, that we are being used. A social network is a good thing; one used for advertising, controlled by a corporation, and used to spy and surveil users, is not.

No positive theory can be easily offered here; no suggestion of an alternative system can be made. Yet, individuals should be aware of what is happening to them, and how they are changing. That sensitivity, at least, should help them navigate these new, uncharted waters of communications and relationships, and ultimately how they see the world, and themselves in it.

Individuals should, above all, realize that they are being trained. Nowadays, there is a quite often mention of "machine learning," of how "training" the machines with large data sets can make them become smarter, better thinkers, more adept at solving problems. Yet, human beings do not seem to consider that the machines and interfaces they interact with are training themselves indeed. Modes of communication force personal communiques into the formats they require and permit; individuals are learning to express themselves in Facebook statuses and tweets. They are becoming different beings as their relationship with our informational environment is changing.

As mentioned earlier, exhibitionism and individualism within our contemporary culture results in a self-imposed slavery. In order to establish a true humane society and a civic life, a right sense of privacy should be established by putting all of our material wealth and technological prowess at the service of something greater than our own material success. Yet, often times this something greater is being confused with a greater, better thing, technology, or trend out there in the material world. Nowadays, this something greater seems to be the artificial intelligence (AI), which will be covered in more detail in the next section.

2.1.1 Thinking Machines

The notion of "thinking machines" and the field of AI lead to many interesting philosophical questions. Artificial Intelligence refers to the notion of developing computers that are able to think and perform tasks that normally require human intelligence. While AI arose as a possibility quite early in the history of computing, the concept of a machine that could mimic a human has a much longer history, going back hundreds of years to the idea of automatons. Modern debates on AI proceed on the presumption that we will remain static while machines continue to change and "take over us." Yet, as technologists, we shall not be waiting for the rapture of the geeks. Instead, our understanding of humanity, technology, and the future should be shaped by a knowledge of God who made us his stewards on Earth. As God, the Almighty, mentions in the Qur'an:

> *"And it is He (God) who has made you successors (khala'ifa) upon the earth and has raised some of you above others in degrees [of rank] that He may try you through what He has given you. Indeed, your Lord is swift in penalty; but indeed, He is Forgiving and Merciful." [6:165]*

In the end it is God who will one day make us return to Him, not in the virtual world of a computer, but in the Afterlife.

If the world we live in is one that our machines will be able to "take over," that world will be unrecognizable to us, because we will be unrecognizable to our present selves. Part of that transformation will come about because we will have been trained to think, read, and write differently by the machines; these machines and their technology, the systems, the rules and laws, and techniques that sustain them are constitutive aspects of ourselves and our societies; radical changes in them induce radical changes in us.

We are used to looking at older photos and exclaiming in surprise and wonder at how much we have changed; those photographs have never captured the changes in our interiority. But a history of our social media interactions most certainly will; we might be surprised to see what we are becoming and have already become.

Even though we may easily feel God's creative intelligence when observing a beautiful natural surrounding such as a lake or a forest, we may never experience the same feeling when using an innovative technology or latest Smartphone. Why is this the case?

The answer lies partially in the fact that the mindset of the modern man has been used to contribute the successes of technological progress as approval of self-sufficiency and individual power as a result of the scientific materialist worldview. Despite the efforts of Islamic scholars in the past to establish the seedbed for the cultivation of scientific revolution in the West, by providing both a moral and an intellectual framework in order to investigate the created order, most of us are inclined to view modern technology and science somewhere between being spiritually irrelevant and atheistic as if they have no Islamic origins at all.

Given the fact that a mobile phone could contribute to the sustenance of the created order – in the form of maintaining or strengthening ties between friends and families, could we not re-conceptualize the cell phone as a pointer toward God? This may appear counterintuitive to most of us, yet there are various reasons to make such claims. Given the ubiquitous digital artifacts of our own time, these can be considered as reminders of the goodness of the Creator in the form of pointers toward not only His invisible power but also His care and provision for humankind.

Moreover, a more profound point should be raised with regard to God's glory as well. While we may be dazzled by the amazing harmony among the parts of a flying bird, we should be more inspired at the creativity necessary to create human beings with the intelligence to develop the Concorde. Although engineers can bridge great seas and scale huge mountains as modern craftsmen, only the Master Craftsman can create human beings who can achieve these goals.

If we can achieve a subtle shift of our perspective, we can start digital artifacts such as mobile phones and contact computers as being powerful pointers toward the Divine rather than being merely as physical objects devoid of any spiritual significance.

According to a recent survey in the field of AI, given the advanced state of this field, today's autonomous vehicles can resolve the challenging problem of off-road driving. When it comes to analyzing certain kinds of technical images, (e.g., in medical applications), computers seem to do well or better than human beings.

Despite these notable examples, AI still faces many fundamental challenges. Even for simple narrowly defined tasks, AI still generally lags behind human abilities. It was also observed that AI capabilities often reach a plateau and that any incremental improvements typically require tremendous efforts and computing power. One area that presents particular challenges for AI is the area of common sense reasoning. A category of questions called "Winograd schemas" can be used to test such reasoning. An example is the following: "The man couldn't lift his son because he was so heavy. Who was heavy?" Statements like these simply require identifying who the pronoun "he" refers to, yet they rely on broader knowledge to infer that heavy items are more difficult to lift. Nevertheless, researchers remain busy trying to tackle problems. Could we ethically turn it off or destroy it, saying it is simply a computer and therefore we can do what we wish with it? What role does humility play in considering such a technical marvel – or perhaps monstrosity? Does humility say humans should never dare to develop such devices?

Moreover, the growing complexity of AI software presents numerous challenges, especially when it is used to control automobiles, surgical robots, and weapon systems. It is a particular challenge to verify systems that rely on "machine learning" techniques. Apart from the risk potential of weapons of mass destruction, there is the risk potential of destructiveness of knowledge-enabled mass destruction (KMD), as they can be amplified to a great extent by the power of self-replication. According to a BBC interview with Stephen Hawking, such a development of a complete version of AI could bring the human race to an end. Similarly, the engineer and entrepreneur Elon Musk asserts that human beings should treat AI very carefully as it may pose the biggest threat to their existence.

The Internet has helped entrepreneurs slim down the scope of their firms, instead facilitating peer-to-peer connections or commerce. When a business like online retailer Amazon operates with a deep commitment to all of its customers, enormous opportunities are created. Netflix, the popular video streaming service, is a significant business customer of Amazon Web Services, the cloud computing platform powering the e-commerce giant and available as a service to startups and Fortune 500 corporations alike.

The speed of communication has aided growth but the most successful marketplace businesses have developed ways of signaling the reputation and integrity of buyers and sellers. After an Uber ride, the driver and passenger both rate each other with consequences for their future access to the network and service. Similarly, AirBNB and other vacation property rental market-places encourage the host and guest to both provide feedback on their stay. New businesses are providing transparency and aggregating reputation so customers and producers can make better decisions (e.g., TripAdvisor, Yelp, Angie's List). This is both transactional (Where should I stay on my trip?) but also character forming (How can our team better serve others?). Third parties are also building on top of this reputational ecosystem. For example, the lender OnDeck Capital has incorporated Yelp reviews and similar data when underwriting loans for small businesses.

The business model for many services relies on revenue from targeted advertisements that require attracting as many eyeballs as possible and keeping them coming back. These nudges have become increasingly sophisticated, with some social media companies hiring behavioral scientists to help advise developers on tuning apps so they play on a user's dopamine levels.

Jeffrey Hammerbacher, an early social media pioneer, once lamented in an interview with Businessweek magazine, "The best minds of my generation are thinking about how to make people click ads." Some companies have recently been formed to provide technical solutions to the distractions that arise from digital technology. Flipd is a software company that has created an app to help people spend less time on their phones and remove distractions. The app has been adopted in universities to monitor students' smartphones and encourage them to remain focused during classes and lectures. Another program called SelfControl helps users block distracting websites while working on their computers. While technical solutions can provide helpful aids, the problem is not just with our time and our eyeballs – it gets to our hearts. For many our screens are continually with us – when we rise up, when we lie down, and when we walk along the way.

Human beings can utilize their capacities for good or evil which is also true of their capacity to develop tools or technologies which can be directed toward noble or unworthy ends. Ultimately, it is up to the developer of these tools or technologies whether he puts his creativity, freedom, and intelligence for devising a tool or technology toward evilness or goodness. An in-depth understanding of the potential of technology to be utilized toward good or evil should be in fact one of the main concerns of the innovators of the future. Could a religious Bill Gates or Mark Zuckerberg develop a new trajectory for a humanitarian innovation in the field of technology? Is there a way for articulating one's technological innovation with a spiritual depth and faith dimension while standing at the forefront of new technologies which shape the human life in profound ways? Can new technologies such as quantum computing, organ printing or robotics and augmented reality be utilized to alleviate human suffering? How can mosques utilize Google Glass or similar other tools of augmented reality creatively in order to deepen the human worship or to provide training to outsiders on the Qur'an and Islam on major topics such as doing the pilgrimage or paying zakat to create a greater inter-religious understanding? The new generation of Muslim technologists should ponder these questions and many others that cannot even be anticipated now.

Although it was Hitler and his henchmen who unleashed death and destruction during the World War II, someone had to design the railways, factories, warehouses, and machinery for their war effort. An article in the *New Atlantis* titled "The Architecture of Evil" includes the provocative statement that the furnaces of the Nazi death camps have been designed by some individuals. The article goes on to describe the life and work of Albert Speer, Adolf Hitler's "chief architect," reminding us that Hitler did not work alone. The truth is that engineers and architects designed the technology that enabled the Nazi brutality. Speer later wrote that his obsession with output statistics and production made all feelings of humanity and considerations blurry.

"The Architecture of Evil" not only tells the story of Albert Speer, but goes on to suggest that in order for today's engineer students to mature as responsible citizens who can go the extra mile beyond the immediate needs of their technical work, their education should focus on both the liberal arts and analytical skills. Our computer science and engineering schools need to attend to ethics and values if we hope to build a just society.

I would add that Muslim engineers must see their technical work as a response to God, one in which even our mathematical models, computer programs, and architecture need to enhance justice and show mercy as we walk on the straight path to our Creator.

While Speer's situation seems like a dramatic example, the truth is that all engineering work involves some moral choices and responsibility. Even programmers writing logical, mathematical code need to recognize that their creations are not neutral and unbiased. Cathy O'Neil worked as a math professor until 2007 when a lucrative opportunity arose to use her Ph.D. in mathematics at a hedge fund. Shortly thereafter, the financial crisis occurred and O'Neil found herself pondering her work and her role as a "quantitative analyst" (or what is often referred to as a "quant") in the finance industry. Reflecting on this, she later asserted how the collapse of major financial institutions, the housing crisis, and the rise of unemployment are also within the responsibilities of mathematicians who were trying to wield magic formulas. Her disillusionment led her to participate in the "occupy Wall Street" movement and eventually write her phenomenal book *"Weapons of Math Destruction: How Big Data Increases Inequality and Threatens Democracy."*

In her excellent work, "Weapons of Math Destruction," Cathy O'Neil provides details on why algorithms should not be considered as being neutral and how they reproduce human biases and agendas. We cannot build a neutral platform, every decision matters, including the decision to collect certain types of data and not others. Our collection, analysis, and use of data may even be informed by particular ideological and political agendas.

Programming is not a detail that can be left to "technicians" under the false pretense that their choices will be "scientifically neutral." Societies are too complex: algorithms are political. So how much time do those with technological power spend on exploring the ethics of what we do?

The inner working of these systems is often opaque, and the verdicts are accepted without question. Often, the inner assumptions and values embedded in the math are hidden over concerns of intellectual property and trade secrets. This leaves those affected without any explanation or recourse for unfair decisions which may impact them.

Cathy O'Neil suggests that tech giants are solving problems not relevant to the everyday person and that possibly sensitivity training might help. I feel that something like this would not be enough. No one would argue if some elements of humanity or ethics would be infused into computer science curricula. Even human-centered design does not go far enough. Based on my doctorate research experience, I can suggest that designers of technology consider more "empathetic and participatory design" where there is some degree of involving people who are not in the technology company as autonomous persons in product design decisions, and not just using them as research/testing subjects. So, perhaps, we should design, fund, and celebrate more programs that promote humanity and social justice rather than technical abilities. We should not be targeting creating more factory workers. We should be working on the values of factory owners and managers.

There is little direction of working with software engineers and programmers to help them think in more humane and ethical ways about what they're designing, to be more critical and aware of the underlying politics of what they do. Non-programmers can become more critical citizens when they understand how their (digital) lives are influenced by algorithms, but more importantly, shouldn't we care about the critical citizenship of the programmers? After all, it is highly unlikely that an amateur coder will be asked to design the next big neural network: as unlikely as someone with a casual interest in medicine, or who studied holistic medicine, being called on to perform life-threatening surgery. While there has not been enough focus on how to make young individuals appreciate the social consequences of their algorithms and code, we also could not go beyond the rather utopian and naïve ideal that if one can grasp the meaning of an algorithm and learn how to develop one, then he can shape the computer code rather than it shaping himself. In order to understand the real social power of algorithms, different kinds of knowledge are required.

Technology is not neutral. Even equations and computer algorithms, which may initially appear cold and neutral, reflect the values and assumptions of the people and organizations that construct them. "Big data" sifts through vast oceans of data to find patterns that are then used to inform decisions in areas as diverse as finance, banking, hiring, marketing, policing, education, and politics. While mathematical models allow decision-making to be more efficient, they can sometimes hurt the poor, target predatory ads, and discriminate against minorities while serving to make the rich richer.

For example, should data like ZIP codes act as a proxy for creditworthiness, for hiring decisions, or for dating matches? It's not hard to imagine how such decisions could perpetuate a cycle of poverty. Should the data taken from current employee profiles be used to guide future hiring decisions? Such a decision could perpetuate biases reducing diversity in the workplace. Some of the consequences resulting from data and mathematical models are unintentional, but in the words of O'Neil, these mathematical models are "opinions embedded in mathematics."

The Chinese government is developing a new algorithm that will allow them to rank their citizens on a so-called "Social Credit System." The goal of the system is to judge the "trustworthiness" of each of the 1.3 billion residents, ranking citizens based on everything from paying bills on time to consumer purchases to interpersonal relationships.

Not only is this entirely intrusive, but it also serves to diminish a series of personal freedoms. As just one example, citizens will likely begin to self-censor their posts on social media given that a negative post about the government may result in a lower score and subsequent negative impacts.

The system also distorts the free market by changing consumer behavior. Rather than only investigating behavior, such a system shapes it. It "nudges" citizens away from purchases and behaviors the government does not like. If consumers do not have the full freedom and power to choose for themselves, society will continually to drift from following economic law. There will be no supply and demand dictating the point of equilibrium. Demand will be artificial and prices will be arbitrary.

Such a system doesn't just reduce freedom, but it contorts the human person to fit a certain mold. It reduces the human person to an algorithm. Totalitarian systems understand people to be malleable and seek to engineer "the perfect citizen." The human person is seen as a hinge in the machine rather than a contributor to society. When producing a material object, imperfections are undesirable, but humans are not objects and therefore should not be treated as such. Algorithms cannot take into consideration context. The system does not know *why* you didn't pay your bills. It just knows that you didn't. It fails to see the citizen as a human person and instead sees him or her as a number.

The removal of these freedoms and the increasing treatment of human persons as objects isn't unique to China. All over the world there are governments and systems that are heavily involved in the business of monitoring and rating citizens and consumers. There has to be a line drawn between defending our freedoms and removing our freedoms.

One of the respected scholars of the 20th century, C. S. Lewis has served as a professor of Medieval and Renaissance literature at both Oxford and Cambridge Universities for almost 30 years. One reason why Prof Lewis has been doubtful about technological advancement was his idea that an omnicompetent state could utilize technology for the rise of its pervasive tyranny. According to Professor Lewis, although a welfare state may seem to be appealing to many due to the wide scope of human suffering existent around us, caution should still be taken about the purveyors of utopian dreams. Instead, Professor Lewis suggests that the good actions of individuals who strife for overcoming the challenges in a dark world should be promoted the art of living involves in dealing with the immediate evils as much as one can.

The fact is that all of life is religious and that even our technical and mathematical work has moral and ethical implications. Our big data algorithms and mathematical models can be directed in ways that are more obedient or less obedient to God's intents for his world. In fact, as more decisions are informed by number-crunching computers, we will need to make sure that justice and transparency are emphasized.

Pinocchio, Frankenstein, and Pygmalion, all of these, are examples of archetypal stories about distinguishing humans from artificial creatures. This theme is also explored in many science fiction shows and movies, such as the affable Commander Data in *Star Trek*, the replicants in *Blade Runner*, the cyborgs in *Battlestar Galactica*, and the robot boy in the movie *AI*.

God sent his Prophets rather than sending mere information or augmenting the reality with some new set of moral obligations. As the Qur'an says,

> *"Unto Him you all must return: this is, in truth, God's promise-for, behold, He creates [man] in the first instance, and then brings him forth anew to the end that He may reward with equity all who attain to faith and do righteous deeds; whereas for those who are bent on denying the truth there is in store a draught of burning despair and grievous suffering because of their persistent refusal to acknowledge the truth." Qur'an: (10:4)*

Technology that focuses on efficiency in terms of human interaction will not fail us in this world, but it also will make us less human.

According to the Qur'an, the God is relational in the sense that He was speaking to Moses and, ultimately, sending the last beloved Prophet Mohammed. While in other religions, such as Buddhism, nirvana or salvation or wisdom is obtained through information, Islam in addition encourages human beings to follow the examples as set in the Qur'an and by the Prophet Mohammed (s.a.w.w).

Being held morally responsible for an act implies freedom and choice, and AI programs simply follow a program. Thus, sin becomes evident in machines when humans develop and employ them in ways that go against God's intent for his creation. Those that hope AI will somehow enable us to surpass ourselves, producing an intelligent, yet sinless creature, are mistaken. Our redemption does not lie in our technology.

It is believed that the perfectibility of the man can be realized via the means of science, technology, politics, and education. Yet, the paradoxes of our modern civilization grow more and more catastrophic each year. The more the abundance grows, the more the resentment becomes while technological achievements seem to be more dedicated to the task of destruction. In addition to this, with the multiplication of the production, scarcities become paradoxically more endemic. Despite the high level of education of human beings in our century on average, even the mere idea of wisdom seems to have vanished from the world.

Modern society and modern man has only continued down that path at the expense of human freedom – all for the glory and fame of man.

These temporal achievements of science, technology, inventions, and the like also have a divine significance.

As AI and the subsequent technology continue to improve, we needn't be fearful of our own position and power. We are not mere machines, but creative and imaginative human persons fully capable of adapting, mobilizing, innovating our modes of service to be in line with his love and purposes.

The value of being a human is derived from their acts based on serving the Creator rather than as producing merely goods or becoming living machines.

In ancient times, celestial cycles may have dictated the ways of living. For example, according to this belief structure, fertility indicated Aphrodite's approval and lightning bolts were indicative of Zeus' anger.

While the Greeks described a good life as one lived in accordance with virtue, Muslims live under the commands of God's moral law. This is why Muslims pray five times a day in remembering the Creator or fast during the holy month of Ramadan as God's commands. Given the commands of the Creator for its creation in the Qur'an, human beings are re-oriented to develop specific living patterns. Given this dynamic relationship between the Creator and its creation, human beings can expand their horizons in arts and technology with the intention of having a better grasp of the Truth in God's world. Human beings were created not to become mere productivity machines, yet to develop meaningful relationships with both other human beings and their Creator. To ignore the Qur'an would result in an ignorance of the nature of man.

According to an article in the *Newsweek*, the drive to develop a real connection between human brains and machines may have mind-boggling consequences. The question that should be answered is not whether such a connection will have consequences or not, yet whether and how such consequences will improve the current situation of mankind.

3

Spiritual Dimension of Technology

Humans have always had an identity crisis. For much of our recorded history, we have used rather specious definitions of humanness or personhood that granted power to some, while granting few or no rights to others. At times some have thought our gender or the color of our skin formed a key part of that definition. If we have erred too narrowly in the past, do we now risk erring too widely?

Some 23% of Americans, and a higher percentage of Europeans, say they belong to no religion in particular. Although this is the result of a centuries-long retreat from faith, the process may have come full circle.

Secularization happened in four stages, each one with its own century:

- The 17th century was the secularization of knowledge. This referred to knowledge without any dogmatic assumptions in the form of reason and observation, philosophy, and science.
- In the 18th century came the secularization of power with the American Revolution and the First Amendment, and the French Revolution, which emphasized the substantive separation of the state and church.
- The 19th century was the secularization of culture when the museum, and the concert hall, and the art gallery took the place of houses of worship as places where you encountered the sublime.
- The 20th century saw the final secularization, which was the secularization of morality. In the 1960s, throughout the West, the two foundations of the Judeo-Christian ethic, the sanctity of life and the idea that there is such a thing as a sexual ethic involving fidelity and the covenantal nature of marriage disappeared throughout the West.

Having gone through four stages of secularization, there are no more stages to go through, short of complete atomization of society.

Four great institutions of the state, market, technology, and science are not able to provide a response to the following questions which any reflective person may ask throughout his lifetime: who are we, why are we here, how then shall we live?

The science explains to human beings the how part, yet not the *why* part.

Technology empowers individuals, yet it does not explain how to put that power into use.

The market provides individuals with choices, yet it does not guide individuals on which choices to make.

The liberal democratic state may provide a maximum degree of freedom, yet it does not provide any guidance as to how to use that freedom.

Within the light of this information, it is not difficult to see how religion will make a return.

The West is not going forward bravely to the future. It is marching heedlessly to the past. People of all faiths may keep asking whether the West can survive without the faith that created that cherished body of rights, liberties, and civil protections that are now vaguely referred to as "British values," or "European values." The fundamental truth that is at the heart of the Qur'an is that the governors exists to serve God through their people.

Our Islamic culture drew from its faith tradition the understanding that all life is sacred, that human persons should enjoy freedom from arbitrary coercion, that we should serve others by engaging our God-given gifts in service of our personal vocation.

At the core of every civilization, religion has been a major approach, therefore, to ignore religion would be to ignore the legacy of the civilization along with its achievements.

Human nature, as fashioned by the Almighty, yearns for purpose, meaning, and deeply significant relationships. The soul in its loneliness stretches out longingly for the transcendent. Even the technology, which provides a seemingly endless variety of utilities and gratifications, cannot meet that need. Nor is it clear that the form it has taken for the last century can endure without it.

Humans have been augmenting themselves with technology since the beginning. Marshall McLuhan suggests that technology and media are "extensions" of ourselves. Tools are, by definition, an augmentation of some ability, from hammers, which help us pound harder, to telescopes, which help us see further. Sometimes we use technology to repair, such as using a splint to guide the healing of a broken bone. Despite this close relationship with our tools, until recently the line between ourselves and our technology was fairly clear. When the technology becomes an integral part of our bodies, the line is a bit more ambiguous. For a person with an artificial hip or heart, most would agree on which part of the person was human and which part was technology. It gets a bit fuzzier if we modify someone's cells using gene therapy. According to Henk Geertsema, the use and development of technical tools to improve certain functions of the human body does not eliminate the difference between machines and human beings. Yet, how much of ourselves can we replace and still remain human? Although in ancient times the heart might be considered the seat of emotion and central to our humanity, in modern times the artificial heart would not disqualify one as being human. What if I start augmenting the human brain? Would a "brain prosthesis" ever be possible? Nanotechnology enhancements might still be years away, but one could consider cochlear implants for the profoundly deaf to be a forerunner of brain augmentation.

The way in which questions such as these are answered reveals certain philosophical presuppositions about what it means to be human. If one presupposes that the material world is all that exists, a notion referred to as materialism, then it is not a far leap to suggest that people are no different from machines. Such a worldview suggests that our mind, consciousness, and our entire person arise entirely from the physical interactions of particles in the brain, a view sometimes referred to as physicalism.

Physicalism has several significant implications. For one, illnesses of the mind are reducible to an illness of the body which can be treated by pharmaceuticals. Matthew Dickerson, in his book *The Mind and the Machine*, makes the case that physicalism has far-reaching implications for areas such as creativity, heroism, ecology, as well as for reason and science. Furthermore, physicalism suggests that brains, and hence minds, could be entirely simulated in a computer.

At the very beginning of computing history which was around World War II, the main principles in computer science have been uncovered by the famous mathematician Alan Turing who asked in his phenomenal paper **"Computing Machinery and Intelligence,"** the question of whether machines could ever think and if so, whether there would be a way for us to know. In order to find out the answer of this question, Alan proposed a test to specify whether a machine could be described as "thinking." According to this test, an individual was sending messages to both a remote human and a computer while sitting in a room, and was trying to understand which is which. According to Turing, if this individual could not tell the difference between the computer and the human being, then one could classify the "machines thinking without expecting to be contradicted." This famous test has been referred to as the "Turing test."

Turing's early writings were followed by rapid developments in computing and the development of the field of "artificial intelligence" or AI. The notion of "thinking machines," the basis for the field of AI, naturally leads to many philosophical questions such as follows: What is a mind? Will machines be able to think as humans do or can they just calculate? What does it mean to be human? What is the difference between people and machines? Could machines ever have free will?

Tests like the Turing test are based on the faulty presupposition that to mimic a human implies that something is human. As machines are increasingly able to do things that people do, distinguishing humans by the functions they can perform will become less meaningful.

Ray Kurzweil, an accomplished computer scientist and author of several books including *The Age of Spiritual Machines*, suggests that within the present century we will be able to download our brains into a computer and thus escape our mortality. All that remains to achieve this is for neuroscientists to map the brain and for sufficiently powerful computers to be developed. At that point, it is suggested that our brain could be uploaded into a computer and can live forever in this way without being restricted by our mortal bodies. Some have coined this "**the rapture of the geeks**." This vision has inspired **books**, **conferences**, and efforts like that of the **Terasem Movement Foundation**. In the case of the rapture of the geeks, the end goal results in literally becoming like a computer.

The goal of downloading brains into computers is based on a materialist worldview which sees the physical world as all there is, as an independent reality, self-existent and hence divine. In turn, it is the materialist reduction of what it means to be human that allows the rapture of the geeks to be conceivable. This worldview connects a trust in technology, a rejection of God, and a reductionist view of what it means to be human. A knowledge of God as the creator is tied to an understanding that as fallible creatures we need a Supreme Savior.

Such a worldview reveals a trust in technology, one that even promises eternal life. Andy Crouch claims in his latest book the following two simple promises provided by every idol:

- "You shall not surely die."
- "You shall be like God."

The Qur'an includes several verses on the subject of worshipping such idols.

"It has been revealed to you, and to those before you that if you ever commit idol worship, all your works will be nullified, and you will be with the losers." [39:65]

"Indeed, his crops were wiped out, and he ended up sorrowful, lamenting what he had spent on it in vain, as his property lay barren. He finally said, "I wish I never set up my property as a god beside my Lord." [18:42]

Different schools of thought in ontology (the philosophy that explores the nature of being or existence) have suggested anthropologies that affirm or deny the existence of at least three different parts: the body, mind, and soul. The body is composed of our physical self, including our neurons and brain. The mind consists of our thoughts and consciousness. The soul is "that part of us that might be said to be eternal or to transcend in some way the mortal body." Most anthropological views can be categorized as either monism or dualism. Monism asserts that humans are made of one substance. Thomas Hobbes was an early supporter of monism by arguing that consciousness and souls arise from the functions of the body alone. In contrast, dualism holds that humans are somehow made up of two parts, often identified as the body and the soul. Dualism includes many theories about how the body and soul are separate but related. Platonic dualism saw the body as an earthly package for the spirit, something to be eventually discarded. René Descartes, an early modern philosopher who promoted a form of dualism, suggested that the body is like a machine that interacts with the mind. Although the Qur'an is not a philosophical anthropology textbook, there are many verses indicating that we are more than our bodies. God says in the Qur'an:

"And when the souls shall be joined with their bodies." [81:7]

More recently, many modern Western philosophers have embraced materialism, which is a form of monism that denies the presence of a soul and holds that reality is made up of only the physical stuff around us. In his book, The Concept of Mind, Gilbert Ryle rejects dualism and ridicules it as "the myth of the ghost in the machine." Similarly, a materialist view has been promoted by Ray Kurzweil in a series of books such as The Age of Spiritual Machines and How to Create a Mind. This perspective dismisses the notion of a soul, concluding that our mind and consciousness arise entirely from the physical brain. Some materialists account for the complexity of the mind by attributing it to the interactions of many simple entities, like an ant colony.

Although each ant appears to act at random, more complex behavior emerges at the level of the colony.

Our view of human personhood has profound implications. For instance, a materialist view applied to the mind (sometimes referred to as physicalism) will conclude that all illnesses of the mind or spirit are reducible to an illness of the body which can be treated by pharmaceuticals. For a physicalist, being human simply reduces to the interactions of basic particles. However, if we consider ourselves as more than simply a physical body, how does that shape our view of what it means to be human?

Some technologists working in AI have not aimed for replication of humans, but rather for intelligence. Is intelligence, the ability to learn and to apply that learning, an essential quality of humanness? Is it a unique talent of humans alone, unattainable by any other natural or artificial creature? Could a machine have excellent logic and rationality, surpassing humans at deductive reasoning? In 1997, IBM's supercomputer called as Deep Blue could beat the famous world chess champion Kasparov.

Another essential quality of humanness is sentience, which refers to our ability to perceive. Douglas Hofstadter, in his famous 1979 book Gödel, Escher, Bach, explored ideas of recursion, self-reference, and the idea of the "strange loop" as possible layers that might allow the whole to be greater than the total of its parts. This is the idea of emergence that simple components can interact so that a more sophisticated, perhaps intelligent, behavior emerges.

Another essential quality of humanness is emotion, which is often considered a part of our intellect, but a peculiar component that is not logical or calculating, even though it can often be predicted. Emotion seems to be connected both to our state of mind and to our bodies. Emotion makes our hearts race and our hands sweat. It puts the bounce in our step or the frown on our face. In order to feel emotions, one must have both intelligence and self-awareness.

Furthermore, as Muslims, we consider the soul an essential part of our being, in fact, the one part that survives our death. In addition, this is often the attribute that many believe uniquely defines us as humans, particularly when other attributes do not seem sufficiently unique because we find them at least partially in other creatures. The soul seems to be confined to humans. However, one cannot measure for the presence of a soul as a test of humanity. The computer scientist Matthew Dickerson makes the astute point that assuming we can scientifically test for the spiritual assumes that the spiritual is reducible to the material, which is equivalent to saying that the spiritual does not exist.

Rather than distinguishing humans by how they think, as homo sapiens, many point to our ability to make tools as what distinguishes us from other creatures. Thus, we are homo faber, humans as makers. Inventing novel devices, composing new music, and innovating in business are all examples of creativity that may also be hints of an essential quality of our humanity. Humor is a type of creativity required to banter about with one another, and laughter is sometimes considered uniquely human.

Another essential quality of humanness is free will. We cannot be held morally responsible for an act unless we have a choice (to act or not to act). Moral agency, the ability to choose, and to be held morally accountable for our choice, is perhaps uniquely human. While some would argue that a computer can never be human because it cannot truly make a free will choice, others counter that humans cannot make a free choice either, thus subscribing to some version of determinism.

Taking a materialist view to its logical conclusions would deny the very possibility of many of the attributes listed above. As physicalism applies the principles of materialism to the brain, our brains are considered as natural phenomena, and as a result, the laws of cause-and-effect found a manifestation in machines as well. A strict physicalist view would deny the presence of a soul, suggesting we are just bodies operating under physical laws. In addition, it would reject the notion of free will. Furthermore, if our thoughts are merely the "interactions of basic particles" then true creativity is also an illusion. Matthew Dickerson argues that "to the extent that creativity is defined in terms of originality ... physical automata, whether the digital computer variety or the biochemical human variety, are not capable of originating anything."

Materialism and physicalism are highly reductionistic with profound implications on how we view our humanity. Human attributes such as souls, free will, creativity, and emotions are essentially an illusion reducible to the laws of physics. Furthermore, it has implications for our understanding of knowing and truth. Materialism can be considered as a category of naturalism, and according to C. S. Lewis, naturalism provides a complete account of our mental behavior, which does not provide any insights on which the complete value of our thinking, with regard to truth depends, or in other words, there is no room for the acts of knowing. Ironically, a physicalist view of reality even leads to devaluing of our physical bodies, potentially leading to a new kind of gnosticism. Even if one rejects a physicalist view, it is not clear that any of these attributes or any other proposed characteristics can definitively categorize humans and non-humans. Not only is it difficult to conclusively identify those attributes that are sufficient, it is also difficult to simply list which ones are necessary. Perhaps the value of such a list is not as a tool to determine who is in the human "club," but rather to encourage and challenge each other to flourish and grow toward the best humans we can be.

Matthew Dickerson argues that creativity is the ability to make something original and "to bring into being something new, which does not proceed entirely from what has gone before or what already exists." He argues that machines are controlled by physical causes (predictable or unpredictable) and therefore, by this definition, cannot be creative. Esthetic ways of knowing cannot be simply reduced to physical processes. In contrast, a materialist view of humans would suggest that creativity is just an illusion. Regardless of how we understand our status as created beings in the likeness of our Creator, there is a danger to defining a line too tightly around our humanity. If we define ourselves as intelligent beings, then what does that say of people of below average intelligence? If we require emotion or creativity, what does that say of the person lying in a coma? Thus, the danger in circumscribing our humanity too tightly is that we consider certain people as somehow less human, whether they are a fetus, a senile elderly person, a man in a coma, or a child with a severe brain injury. It would be a mistake to define some kind of litmus test for what it means to be human based on attributes such as intelligence or creativity. Our ontological status as humans seems to be distinct from the rest of the creation (including machines). This implies that human personhood needs to be attributed even when certain human attributes are less evident due to age, capacity, or infirmities. A materialist would reject the notion of a soul, free will, and creativity since this viewpoint sees everything as determined by natural laws. An Islamic perspective is shaped by the understanding that God created us with the ability to respond to him, and with that ability comes freedom and responsibility. Freedom and responsibility imply that we have a choice. The ability to choose (and especially to make a moral decision) is perhaps the most difficult attribute to understand about ourselves. How can the creator give the thing he creates the ability to do something other than what the creator intends? And yet that is what God gave us. This ability touches on the paradox between election and free will. This ability makes us human, and perhaps more than any other ability makes us distinct from machines, but this ability also allowed us to fall.

Views such as materialism elevate, leaving aside any type dependence or responsibility to God, one aspect of human being to be the ultimate one. Anthony Hoekema classifies this as a type of idolatry as it makes worshipping an aspect of creation rather than the creator. Others place their trust in the hope that one day we will be able to download our brains into a computer and thereby achieve a kind of immortality. This is an example of technicism, placing our trust in technology as savior of the human condition.

From the early days of computing, entertaining examples of software pretending to be a person have arisen, such as the ELIZA program written in the 1960s that responded in natural language using scripted pattern matching. Today's expert systems can be surprisingly humanlike, such as computerized call center operators that understand a wide variety of spoken phrases, or the IBM Watson supercomputer that can defeat even the best humans on Jeopardy!, the popular TV game show. Even our smartphones are getting smart with us, with Siri providing helpful and sometimes entertaining answers to our queries. The Google autonomous car has demonstrated the ability for intelligent software to navigate a vehicle in complex environments. Web search engines routinely provide relevant results with both accuracy and speed.

Following his work with ELIZA, Joseph Weizenbaum reflected on the appropriate role for computers. He concluded that computers ought not to be used for tasks that require wisdom. Weizenbaum goes on to conclude that "there are limits to what computers ought to be put to do." Our purpose is to love and respect God, to fulfill his commands, to establish justice on this earth and to show mercy toward other human beings. These are things that we ought not offload to machines. Why? Regardless of the question of whether machines could actually do these things, humans ought not delegate those tasks that form our very purpose. Tools that aid us in our purpose are commendable, but tools that purportedly perform our purpose instead of us are condemnable. Imagine inventing a machine that rather than helping us pray or worship, instead does our praying or worshiping for us, so that we no longer feel the need to do it ourselves. Such a machine would be completely misguided, and the users of such a machine would be truly deluded.

We could use AI technology to help us seek justice by enabling an attorney to help less fortunate members of society at a reasonable cost, by using an expert AI system as a first contact "help desk." However, it would be important that the attorney does not simply sit back and let the expert system provide the only advice to the client. Rather, the attorney ought to use the expert system as an assistant to do a first interview, so that her or his in-person follow-up meeting with clients is more effective. Some might argue that manual labor is drudgery and the machines free us to do more creative work. Now with AI, we have a new variation of this dilemma as expert systems are developed in order to replace experts such as doctors, lawyers, or perhaps even engineers. But does freeing us from labor free us from the very activity that makes us human? In his book The Glass Cage, Nicholas Carr provides a nuanced discussion on the many effects of automation illustrating that it is an ethical choice since it shapes our lives and our place in the world.

We could use AI technology to help us love mercy. For example, AI image processing systems already exist today that can detect certain types of breast cancer in mammogram images better than human doctors can alone. As another example of mercy, we could enable caring by providing a first contact for call centers with AI natural voice recognition so that trivial tasks could be completed routinely, while ensuring a human operator smoothly steps in for more creative and service-oriented needs. Home automation systems could enable the elderly to maintain independence longer by assisting them with everyday tasks such as cleaning and meal preparations. However, it is important that such systems do not entirely replace humans. For instance, the design of AI programs and robots should recognize social norms and not be employed to replace human care and companionship. Sherry Turkle observes that any relationship with a robot is a relationship only about one person. We could use AI technology with humility by recognizing our own human limitations. If we are uncertain of the status of our AI creations, then in humility, perhaps we should avoid such pursuits. That is, perhaps there is a line beyond which we may develop technologies for which we no longer fully comprehend the implications. We have a long history of letting the genie out of the bottle, and we know you can never put him back in. This is the risk illustrated in the tale of the "sorcerer's apprentice." However, even if some or even most agreed to be prudent with research and development, a few might continue these developments. As we consider the AI tools we build, we need to keep in mind our purpose. All technology is utilitarian: we develop tools as means toward ends. But technology has a bias, and this bias shapes us as we use our tools. How can AI help us fulfill our purpose without the means distorting our ends? For one thing, we should not aim to develop thinking machines that replace us, but rather to develop thinking machines that aid us in thinking ourselves, that augment and extend our abilities. On the topic of AI, Fred Brooks suggest that we should explore intelligence amplifying software to work together with humans rather than focusing on building "giant brains." In a paper on the benefits and risks of AI, the authors conclude that "Some of the most exciting opportunities ahead for AI bring together the complementary talents of people and computing systems."

The computer scientist Edsger Dijkstra once wrote that the ability of computers to think can be considered as being similar to the ability of submarines to swim. Regardless of where one stands on this question, it is clear that AI raises many fundamental questions about what it means to be human. These questions include issues of philosophical anthropology and the notion of the body, soul, and mind. It includes questions about what makes us uniquely human and whether a machine could ever replicate that. Many attributes are associated with being human such as intelligence, emotion, creativity, and free will. Views that suggest that computers can completely replicate humans are largely based on a materialistic view of humans. The implications of materialism lead to a denial of many of these attributes, such as free will, creativity, and the soul. In fact, materialism can lead to a rejection of the body as people seek to shed their mortal bodies and look forward to downloading their brains into virtual environments. Instead, there is another view of what it means to be human and why we are here on Earth based on the Qur'an. The Qur'an describes who we are as stewards created by God who have been granted freedom and responsibility. We are called to do good deeds also by participating in the use of AI and technology. The question of how should Muslims think about thinking machines is not just an academic exercise nor is it just fodder for science fiction movies. This question leads to fundamental beliefs about what it means to be human. As we better understand a Qur'anic view of ourselves, we will also better understand our relationship to our machines and technology. We are called to use AI in responsible ways that lead to human flourishing and exercising humility to avoid possible harm. We also recognize that there are some things for which AI ought not to be used and which may require limits. Differing philosophical presuppositions lead to very different conclusions about the place and use of AI. These technologies are not neutral, not only in the presuppositions behind them, but also in their increasing impact on our work, our culture, and our world. In humility and in recognition of our capability to commit sin, we should aim to develop tools that ameliorate the effects of sin, enhance justice, and show mercy. In short, our tools should aid us in working as redemptive agents.

From a faith-based perspective, there are several things that provide value to the work of the technologist. Considering the potential of technology to empower relationships among the community members, they can heal the broken world and restore the capacity of human beings to do good. Techno-logical developments not only can provide means to eliminate diseases via vaccines, but also can minimize human suffering.

Dr. Edward W. Younkins asserts in one of his articles on human flourishing that abilities, potentialities, virtues, and individual human potentialities should all be put into rational use in order to pursue one's own freely and rationally chosen values and goals. An action is conceptualized as being right if it results in the flourishing of the person undertaking the action. As self-actualization requires the fulfillment of individual capacities based on a moral accomplishment it can also be considered as moral growth.

Moreover, certain social, legal, political, and economic conditions are required in order for human flourishing to occur as each individual life should be protected and justice should be maintained. The freedom to make decisions is an essential aspect for the achievement of individual flourishing. A mere focus on "justice" and "peace" would not suffice to make long-term plans about a truly just and free society.

It would be a mistake to conceptualize human beings as only being happy or feeling fully alive. The following relationships are important for flourishing:

- God and self,
- Others and self,
- Self and self.

As it is said in the Quran,

> "It is He who has appointed you vicegerent on the earth..." (Quran 6:165).

Indeed, the Muslim's character (khulq) is one that is to be inclined to moderation and conservation rather than excess and wastefulness. The role of human beings in general, and Muslims by extension even more so, is stressed in seven Quranic verses that tie stewardship (khalifa) to the responsibility of human beings to carry out this trust (amana). The Prophet (peace be upon him) said,

> "The world is beautiful and verdant, and verily God, the exalted, has made you His stewards in it, and He sees how you acquit yourselves" (Saheeh Muslim).

Given this crucial relationship between God and self in addition to the others and self, flourishing should be described neither as collectivism nor as raging individualism. Flourishing starts with the Creator, entails personal responsibility, and also has an emphasis on the common good. Strikingly, in order to establish a community based on faith, what is required is a bond of the human heart rather than space, time, matter, or any other "chance" relationship in the world. As this is an act of Divine Grace, no amount of earthly means or human will can bring it about. Allah says in the Qur'an:

> "...And remember the favor of Allah upon you – when you were enemies and He brought your hearts together and you became, by His favor, brothers..." [Quran 3:103]

Therefore, only the Creator alone can bring together the hearts of individuals and binds them.

Among the noblest of means of human flourishing would be the realization of final goals at worshipping God as the ultimate end. Human beings' highest end is to glorify God.

Unless human flourishing takes as its starting point Creator Himself, it may end up with mere self-fulfillment.

In addition to making God the center of everything we do, we should also take the right to dignity; the rule of law; wealth creation; the role of the family along with the principles in the Islamic tradition like serving the poor and doing justice in order to realize flourishing.

Given the proliferation of technologies into our daily lives, can humanity get through the current period of upheaval and economic malaise and enter a "new age" of broad economic growth? Millennials are especially concerned with their sense of work and its meaningfulness. In an age when the tools of economics including social media are too often seen as mechanisms for materialism, faith should foster far more than mere theology by infusing meaning through gratitude and unleashing power through creativity.

The dramatically powerful technologies of Wall Street, Silicon Valley, and Industry 4.0 have provoked, in effect, a worldwide economic revolution, starting in the 1970s, challenging the equally powerful technologies of the automobiles, and mass production.

During the golden age of mass production, in the 1950s and the early 1960s, the interests of business and society converged. Government support for education and health services freed up discretionary cash for people to spend on consumer products. High demand for these products created conditions for growth and profit. It was a robust positive-sum game, a super win–win between business and the majority of the population, resulting in good profits and decent livelihoods.

Then, in the 1970s, the mass production revolution hit a maturity ceiling. New products were less viable; productivity fell; markets were saturated. The welfare state became unsustainable, and national solidarity broke down.

Since then, many businesses have seen their cost advantage and their customer demand migrate abroad, away from their home countries. Low salaries no longer harm business as in the past, so living standards have been declining for decades. This, together with unemployment from offshoring, goes far in explaining the Brexit referendum and the fervor of the US elections in 2016.

The same technology could still lead to an open economy, where the means of production are more distributed. We would then see, for example, decentralization in energy, with micro-production of power, maybe blockchain-based micro-transactions for energy trades, crowdsourced finance, 3D printing, and more innovative means of local food production.

The last time a period of crisis ended, after World War II, there was a concerted effort by many government and business leaders to create a unified, prosperous, long-lasting recovery. The Marshall Plan, the Bretton Woods Agreement, the conversion of wartime industries to peace, and the rebuilding of Europe and Japan all played a role. Unfortunately, today's leaders haven't yet taken on the role they played at this point in past surges. Their stepping up last time was a catalyst for ending the crisis.

We may be in the midst of a great surge of technological and economic change since the Industrial Revolution. The last one, the age of oil, automobiles, and mass production, lasted most of the 20th century and still shapes many people's attitudes. Our current surge started around 1970 and has rolled out information and communications technology (ICT) around the world: It is the age of the computer and the Internet.

Even after 40 years, the ICT revolution is far from complete. It hasn't fully changed our way of life, as previous technological revolutions had done. It has brought a dangerous political shift, the separation of the interests of major global corporations from the interests of the national societies where they are based.

It may be useful to have a general look at the history of computer revolution to see how long it took for industries to be transformed via means of software: Since the invention of the microprocessor four decades have passed. It has been more than two decades since the peak of the 1990s dot-com bubble. The number of broadband Internet users is almost two billion. Given these Internet-based services, new global startups empowered by software could be launched easily in various industries without an investment in new infrastructure and training of new employees, necessarily.

Many of these Silicon Valley-style entrepreneurial technology companies already started to disrupt existing economies. Given the low-level cost of startup development, it would be no surprise that the global economy would become mostly, if not completely, digitally wired for the first time in history.

A factor that may intensify those tensions is the nature of today's technology. We have an amazing arsenal of innovations on the threshold of realization: synthetic biology, quantum computing, blockchain, drones, autonomous vehicles, and private-citizen space travel.

Potential breakthroughs are dangled before us. But as Kentaro Toyama, the former Microsoft research director, asserts, technology should not be seen as the response to the problems of our civilization as rather than substituting for deficiencies, information technology amplified the capacity and intent stakeholders.

The key question is the intent with which we deploy this new arsenal of technologies. In a capitalist economy, there are two critical issues:

- Does an endeavor aim to increase productivity, and thus create wealth?
- Does it then aim to distribute that wealth among many, rather than concentrate it among few?

The Qur'an talks about wealth in the following three ways:

- Hoarding of wealth, which is condemned,
- Sharing of wealth, which is encouraged,
- Creation of wealth, which is often ignored and misunderstood.

Using the metaphor of mining let me mention three "goldmines" that I have sought to dig into by writing this book.

- The Qur'anic goldmine: As God created a world that flourishes within diversity and abundance, He also created human beings to both create products and services for the common good and serve Him and His creation. In short, service is a God-given gift.
- The historical goldmine leading to transformation is not new: Individuals and nations can be lifted out of poverty through means of wealth creation which can eventually result in a holistic transformation. This kind of history needs to be further explored.
- The global goldmine: Wealth creation is not a Western or rich-world phenomenon. Many men and women are making a difference through businesses on all continents. Wealth creators should also be enabled to provide service among other peoples and nations in the marketplace. We need to learn from them and others and to extract the global gold.

Despite the universal call to generosity and contentment, material simplicity is a personal choice. As the aim of wealth generation through business goes beyond giving generously, good business can become an agent of positive transformation within the society given its intrinsic value as a means of material provision. As God commands in the Qur'an:

> *"O Ye who believe! Eat not up your property among yourselves unduly. Let it be trade amongst you by mutual agreement."*

As Islam does not approve of a life of asceticism and abstinence, it encourages the undertaking of economic activities to a great extent. Wealth is classified as being among a blessing of the Creator and seeking wealth in a proper way is even considered by some schools of thought as an act of worship. Wealth creation is a godly gift. Various material blessings can and should result from its proper use and be beneficial to the greater community. Businesses can contribute to positive, holistic, and redemptive purposes. Business provides not only a unique capacity to develop some financial capital, but also the ability to develop different kinds of wealth for many stakeholders, including social, intellectual, spiritual, and physical wealth. However, the issue of wealth creation is too often neglected or misunderstood. One major stumbling block is the sacred/secular divide. We need to see that God's concerns are holistic, and so is the mission of the mosque. The lack of business experience and exposure to business among imams and the lack of relevant teaching on wealth creation are among the reasons that so many Muslims hear little teaching, preaching, or discussion in the mosque about the activities that engage the greatest proportion of their time in between times of worship – that is, their daily work. Other reasons include the perceptions of corruption in the business world and the lack of visible structures for commissioning and sending. Some steps can be taken to address these obstacles and to engage the mosque in equipping business people to serve in the marketplace:

1. *Enlighten*: To create awareness through conferences and other means.
2. *Educate*: To accomplish a shift in the minds of people from interest to commitment.
3. *Equip*: To serve as a boot camp for aspiring individuals to become missional entrepreneurs.'
4. *Empower*: To design a roadmap for action.

The Mosque can and should be a part of helping individuals, communities, and economic and social structures work toward a state of comprehensive flourishing, with the elimination of both economic and spiritual poverty, to the glory of God.

"O you who have believed, indeed many of the scholars and the monks devour the wealth of people unjustly and avert [them] from the way of Allah. And those who hoard gold and silver and spend it not in the way of Allah – give them tidings of a painful punishment." (9:34)

In the Qur'an, God is described as the ultimate source of wealth, even when that wealth is produced through the economic activities of individuals. But just as often God is described as one who expresses his judgment in stripping wealth away. God is the author of both wealth creation and wealth destruction as mentioned in the following verse:

"and to whom I have granted resources vast," – 74:12

Even though God is the giver of wealth, human agency plays a part – whether good, bad, or neutral in a moral sense. Sometimes, wealth may arise through wisdom and understanding and sometimes through skill in trading. But sometimes wealth can be amassed through immoral means such as the oppression of the poor, collection of exorbitant interest, or injustice. There are additional kinds of wealth as well as poverty, beyond the material. In addition to the physical and financial abundance, the Qur'an also refers to God's blessings, in examples of generosity, and in warnings against greed and against misdirection of the heart.

Wealth creation basically refers to economic activities that generate profit, that increase income, that create jobs, and that enhance material well-being, so that people (and those who are employed by them) are lifted from poverty, and able to flourish. Yet, we also have in view God's goals for holistic well-being which goes far beyond "peace" in the sense of the absence of conflict, but which describes a state of comprehensive well-being – economic, psychological, ecological, relational, and spiritual. As businesses can contribute to these positive and holistic purposes, they can and should assess their activities in terms of a fourfold bottom line – financial, social, spiritual, and environmental.

The new technology giants, like Google, Facebook, and Apple, along with others developing robotics and similar technologies, will presumably comprise the highest productivity sectors. Yet, they won't lead us to a more decent society unless they encourage distribution. Otherwise, they are unacceptable monopolies. It's not just redistributing income that's needed, but also fostering multiple novel job-creating activities, which historically have been associated with changes in lifestyles.

Given the rapid rise of RFID tags, DNA profiling and analysis technology, national identification cards or identity theft techniques, and, needless to say our own identities should be closely protected. We need to ponder more in-depth on how much of individual lives should interact with the Internet which should only be conceptualized and put into use within that regard rather than being seen as the only source of social interaction or entertainment.

In the post-industrial Information age, the interaction of culture and religion focused especially on new forms of technology and communication which raised important issues for the ways in which Muslims communicate with each other and with non-Muslims.

The basic question also gave rise to other concerns such as how the Islamic message will be affected by the new media or whether and how Islamic discourse may be affected by a spirit of divisiveness.

The Creator can be glorified in the new media. The way of engagement is in appropriate alignment with the divine instruction. Some simple rules that should be taken into account for a theological ethic for Internet discourse can be summarized as follows:

- The "theology" part: At its most basic level, theology refers to the language about God; therefore, when speaking about God, in an academic or popular way on the Internet, one is "doing" theology.
- The "discourse" part: This concerns the interpersonal communication which avoids one from lying, slander, and the like, while actively advancing our neighbor's good name.

Similarly, in the digital realm, the following elements should constitute a dialog: charity, civility, and humility. We need to continue our conversations with fellow Muslims and non-believers in a way that is oriented toward loving them for the sake of the Creator.

This should not mean that our disagreements should cease in favor of a post-modern "Can't we all just get along?" mentality or that we should simply try to "win" the dispute. Rather, what it means is that despite the sharp level of disagreements, rhetorical techniques that aim to reconcile the dignity of the other person should be used. Allowing nothing but God's Word to stand between us, judging and helping, goes beyond the service of mercy as it is an ultimate offer of genuine community.

It should also be taken into account that a civil online discourse goes beyond niceness and mere etiquette as it requires also the dignity with which the other person is treated online. There can be both doctrinal clarity and a humble and generous discourse.

Despite having spent so much time on trying to promote civility within the digital realm, I still get a little nervous on this task as it takes constant effort not to err on one's side. I would hate to support the cause of a freewheeling sense of divine generosity that does not establish vigilance in defending the truth of the Qur'an. By contributing to the truth in love and civility within the digital realm in a digital age, we would realize that humility is a great way to encourage manifestation of these qualities, if we aim to continue our path to the truth in a digital age.

4

Communicational Dimension of Technology

In case you have not examined your beliefs about social media and technology as a Muslim yet, it may be time to reflect on the spiritual implications of using technology with regard to the development of our individual faith.

Although some of us may consider that given the technologies which make our lives much easier, we may be living in incredible times, others may feel not so lucky as there is not so much time to process one progress after another. Perhaps, an approach of numbness or resistance can be taken or we can choose not to be as thoughtful as we should be.

Technology can be considered as a culture-making activity which involves an exercise of freedom and responsibility by human beings. In addition to their intended use, digital artifacts can also shape our world in economic, social, and cultural ways. As the famous aphorism "The medium is the message" of the media scholar Marshall McLuhan states, new ways of living and communicating are delivered by means of technology in addition to the content. To give a more specific example, the automobile is more than a "medium" for getting from point A to point B; it also has a "message." The automobile brings with it personal autonomy and allows us to separate where we live, work or shop, or even worship. Most mosques are surrounded by a large parking space. Many individuals in the Jummah prayer do no longer go to the same mosque in the same local neighborhood where their office or house is located as their cars enable them to drive to other mosques of their own preference as well. No longer restricted by geography, individuals can choose to worship in settings that suit their own individual preferences. As it turns out, our tools not only shape us, but they also shape our faith and places of worship.

Another technology that shaped worship is the microphone as along with sound boards and amplifiers it is at the core of the prayers in mega mosques. Another technology that shapes our worship is the data projector used in some mosques to display Qur'anic scriptures and hadiths. The projector acts not only a means of displaying content, but it may lead to other changes as well in the sense of displaying the Qur'anic verses on screens so that worshippers may be less inclined to look at the physical Qur'an.

It is no longer a surprise that the cyberspace is also providing new possibilities for communities who worship. To give a specific example, "Second Life," being one of the most popular virtual worlds, enables individuals to immerse themselves in computer-simulated environments. In his book *God and Gadgets* Brad Kallenberg asserts that all of the three conditions, namely, time, place, and bodies are required in order for human communication to occur. These refer to the things that technology can make us think as if we can ignore them. Such way of thinking can result in a new form of Gnosticism – if taken to extreme – which reduces the importance of physical bodies in physical community.

With regard to these new tools and technologies surrounding our mosques we should consider fundamental questions about the purpose of worship and the mosque. Although there may be no harm in making use of the best cultural and technological gifts available in our mosques, we may also be wise in considering the following warning of McLuhan: "the 'content' of a medium is like the juicy piece of meat carried by the burglar to distract the watchdog of the mind." So we should better watch against any distractions and ignoring the messages embedded in various technologies.

Marva Dawn suggests that "Worship centers on recognizing that 'great is the Lord, and greatly to be praised' and on responding to that worthiness by gathering with others to praise God as is His due." The media and technology of worship may change, but the message and purpose should remain the same.

Needless to say, worship covers all aspects of our daily life, including our digital activities. As technology is a means for the glory of mankind, we should be mindful to shape them in ways that glorify our Creator.

Augmented reality seems to be now breaking into the world of consumer gadgetry even though, once, it was originally developed for military applications.

On the other hand, we should take into account that as human beings what we yearn for is relationships rather than robots. Rather than looking for solutions via means of programs or computers, we should be seeking prayers. As the Islamic worldview explains the origin of this desire very well, we are created as servants of God on earth.

When it comes to human relationships, we may be wise not looking for efficiency in the first place. Approaching individuals based on a criterion whether they would be worth one's time would be inhumane. In fact, such an assumption well reflected in Neil Postman's famous concept of "technopoly," which refers to the notion that technology can solve every human problem and is hence the king. According to Balaban's explanation, what we need is more information. Yet, what most people are seeking are deeper relationships, rather than more information.

A few years ago Microsoft ran a commercial for the Windows Phone 7. It portrayed a variety of people immersed in their smartphones while oblivious to their surroundings. The series of vignettes includes a distracted father sitting on a see-saw leaving his daughter stranded high on the other end, a hapless man absorbed in his smartphone while his legs dangle in shark-infested water, and a bride walking down the aisle while engrossed in her phone. The commercial is not only amusing, it rings true – many of us are frequently distracted by our glowing rectangles. Ironically the commercial was promoting a new smartphone (albeit one that promised to "get you in and out and back to life" quicker). It's understandable that our digital devices are compelling – they instantly connect us to our friends and family, help us navigate roads and cities, and also provide news and entertainment. Digital technology has brought blessings in education and medicine, and has helped spread the gospel message to the ends of the earth. But many decades ago, prophetic voices like Marshall McLuhan and Neil Postman warned us that while we may shape our devices, our devices also shape us. According to Nicholas Carr, who is the writer of the famous article "Is Google Making Us Stupid?," the Internet keeps diminishing our ability to concentrate and contemplate. Carr claims that our minds are being re-wired to take in information in a swiftly moving stream of particles as the Web distributes it. He makes the analogy of converting from being a scuba diver in the sea of words to a Jet Ski driver zipping along the surface.

More recent research in neural science has demonstrated that our brains remain malleable and change in response to the things we do. Our techno-logical activities gradually shape and sculpt our brains in particular ways. According to the neuroscientist and author of the book "iBrain" Gary Small, our high-technology revolution put us into a continuous stream of partial attention spans in which we can no longer contemplate on thoughtful deci-sions." A thousand years ago, St. Augustine already had an inkling of this when he observed that those habits that are not resisted may soon turn into necessities.

It turns out that the alluring draw of many of our apps may not simply be an unintentional side-effect of pleasing esthetic design. In his recent book The Tech-Wise Family (Baker, 2017), Andy Crouch observes that the "makers of technological devices have become absolute masters at the nudge." These "nudges" come in the form of alerts and notifications which reward us each time we are distracted.

Canadian philosopher James K. A. Smith suggests in his book "You Are What You Love" that our habits and rituals are like liturgies that point our heart in a particular direction. Our digital devices foster habits that have significant shaping implications. Smith suggests that we need to look beyond what we are looking at on our smartphones and consider the rituals underpinned by an egocentric vision which makes the individual the center of the universe. The Qur'an is full of verses that warn us to guard your heart, for everything we do flows from it. The heart is of strategic importance – once it is captured everything else follows along. If habits shape the heart, Muslims ought to devote significant attention to their habits and rituals. Crouch's book "The TechWise Family" provides some practical guidance for "putting technology in its proper place" by outlining "Ten Tech-Wise Commitments" for a healthier family life with technology.

These commitments are meant to nudge us in a better direction and include advice such as remembering the "rhythm of work and rest," and to "show up in person for the big events of life."

Crouch does not merely prescribe countermeasures, but in several sections of candid honesty he shares how his own family has struggled with keeping technology in its proper place.

The Canadian author Arthur Boers, in his book Living Into Focus: Choosing What Matters in an Age of Distraction (Brazos, 2012), builds on the writings of the philosopher Albert Borgmann by suggesting what we need to cultivate are "focal practices." These practices include things like family meals, cooking, gardening, and hiking. He suggests focal practices are things that take time and effort, connect us widely and deeply, and remind us about things that matter. In addition to such focal practices, Islamic traditions offer a rich set of historical practices we can retrieve. Perhaps the best antidote to the challenges of the modern digital age lie in rediscovering the practices found in ancient spiritual disciplines. These include things like reading the Qur'an and the hadiths, prayer, fasting, and serving others. We need to resolve to periodically stem the stream of nudges from our digital devices to re-center ourselves around Islam. Establishing daily and weekly rhythms where digital devices are intentionally and regularly set aside will serve to keep technology in its proper place. It will also have the benefit of helping us avoid the silly situations portrayed in the Microsoft phone commercial. But more importantly, if habits shape the heart, then these habits will serve to point our hearts back to God.

Given proliferation of the technology into our lives, will the new generation be able to accept the idea of moral standards based on the Truth as conveyed in Qur'an? Will technology dominate the lives of the new generation to the extent that it will control their time? Let's further explore some of these concerns how we might respond.

1) *Defining beauty, truth, and goodness*: Due to the predominance of the new digital media, we are exposed to an unregulated and unexamined landscape of ethical concepts which makes the determination of truth more difficult. Ascertaining what is good, beautiful, or true has never been so difficult given the possibility for endless edits on Photoshop, circulated rumors about one's personal life, and any type of self-exhibition on social media. As the new generation increasingly struggles with the idea of absolute truth, how do we know what is true and how is this related to technology?

2) *Fragmented reality*: Despite so many individual experiences in the online world, there is so little coherence which makes it more difficult to bring all pieces together to make sense of reality. Given the high volume of unassimilated and disorganized information, the need to help individuals with foundational frameworks is more important than ever not to mention the increasing level of religious illiteracy regardless of their faith. Helping individuals see where they fit into the big picture of God's story through history is so crucial. Coherent learning experiences are critical gifts.

3) *Approaching technology as a means to paradise or an object of desire*: What does "moderation in all things" mean given the daily average use of social media? Can adults support the new generation in examining their own usage? Among some of the effective strategies to regain a meaningful perspective on this are technology "fasts" and "no-tech Sundays."

4) *Sages or servants*? In order for the new generation to make an impact, they must effectively use the tools of our culture while grasping their limitations and lures. If the search for a grand narrative is not being asked, cultural ideas cannot be affected in wise and conscientious ways.

5) *Balancing persistence, speed, motivation*: While technology is often times used to complete many things at once, our attention span has been changed given these new means of rapid communication and production. It is no longer surprising to be undisciplined in a world of distraction. On the other hand, the potential of technology for more engagement, creativity, and prompt

collaboration in new ways which could never have been imagined before should also be not underestimated. As technology, if used correctly, can increase motivation, and thereby extend learning, it provides a meaningful context for personal development.

6) *Communication fluency*: Apart from technology literacy, it seems critical that for both living out their faith and achieving success in this world, youth need to know how to read social clues, how to empathize, and demonstrate respect by seeing others as God's creatures on Earth. Effective use of social media can be an important means to help develop the kinds of appropriate relationships skills that the new generation needs to learn.

The table below provides a summary of the concerns and individual needs discussed so far along with some practical approaches to better nurture the new generation in an age of technology and social media.

Concern/need	Use Approaches
(1) *Defining beauty, goodness, and truth*	Learn to ask critical questions about the habits of mind and heart in order to develop personal reflection on faith.
(2) *Fragmented reality*	Make individuals understand the coherence of creation by means of project-based learning in interdisciplinary fields.
(3) *Approaching digital gadgets as a means to paradise or an object of desire*	Develop useful habits such as realizing mental idols, and following spiritual paths.
(4) *Sages or servants?*	Convey crucial fundamentals of faith by means of inquiry-learning approaches.
(5) *Balancing persistence, speed, motivation*	Establish flow by keeping individuals engaged and motivated. Increase user choice and context, and provide opportunities for sharing work.
(6) *Communication and relationship fluency*	Provide more opportunities for collaboration and valuing of others in community.

These issues can all be resolved within the light of the Qur'an in the best way. Once we identified critical knowledge and skills necessary for competency and creativity by using the tools of this age, what we need to do is to enhance coherence through learning experiences. Technology and social media have great potential to serve humanity in terms of developing the kind of world-transforming individuals that humanity keeps seeking.

5

Occupational Dimension of Technology

Understanding how knowledge shapes people's decision-making is one of the landmark breakthroughs of 20th-century economics. The famous economics scholar Hayek actually won the Nobel Prize due to his discernment of the "knowledge problem."

Let's look at this question in light of the decisions you and I make every day. How do you know what to cook for dinner or which way to take to go to work? Within the context or scarcity or finite resources, these decisions would be realized quite differently from in a world of no scarcity. There is never a full and perfect knowledge.

According to Hayek, this knowledge is scattered and dispersed as no single person has all the complete information. In one of his essays, Hayek referred to this as the central economic problem. This has been referred to by Hayek as the "triple-whammy" knowledge problem:

- Scarcity refers to dispersed knowledge among individuals.
- The information we possess is most of the time contradictory.
- As unique individuals will all act upon the knowledge differently, individual behavior becomes more difficult to predict to this subjective valuation.

Given these particular circumstances, it may sound actually amazing that we can get so much done at all.

Given this feature of knowledge being dispersed or scattered, the beauty is that markets help coordinate this knowledge through a process of voluntary exchange.

From the perspective of a faith-based foundation for an economic way of thinking, the mechanism by which knowledge is dispersed this way can be described as follows:

Consumers can tell producers how much they value items in the market based on how much they are willing to pay. By sending signals about scarcity and subjective value, prices help us cope with the knowledge problem by giving economic actors a way to share exclusive information and make informed choices based on price signals. Within the context of the Western society, value creation has become difficult to be recognized as it is no longer restricted to planting seeds and reaping harvests or bartering at the village market.

Human beings were designed to act in accordance with God's will to do good deeds on Earth. We have been provided with various cognitive skills such as reasoning and cognition to perceive the material world through our senses and evaluate these perceptions. Furthermore, we have been given skills to use the arts to disclose the Creator's beauty as well as to utilize creativity in various fields such as agriculture, manufacturing, and technology to help flourish the world's economy. Economic exchange empowers human beings while enabling the production of things that are required, and science shows various threads of evidence with regard to the patterns and systems of God's world underpinning his character as sustainer, planner, designer, and so on. Complete submission to God (aka Islam) frees us to put into practice virtues such as love, mercy, self-restraint, courage, and justice to God's design for humanity.

The concept of vocation is multifaceted and complex. In contemporary discourse, it is an idea that is perhaps most often connected with work of some kind, usually (but not always) work that involves remuneration. In the educational realm, vocational schooling is associated with skilled trades or manual labor. What a person does for a paycheck is typically described as one's vocation, while what a person does for other reasons – whether personal interest, amusement, or fulfillment of other duties – is understood as one's avocation.

In our days, vocation seems to be often conceptualized as being either worldly or religious – not both. This modern situation is intriguing as our contemporary world leaves room for such special vocations but largely identifies vocation with worldly endeavors. These worldly endeavors, however, are secularized not only in that they are separated from religious institutions but also and more fundamentally in that they are separated from God. To speak of vocation nowadays leaves out the question of the divine *who* – the one calling a person to do something and to be someone. In place of God is the self, the state, the dollar, or some combination thereof.

Amid modern society's more common materialistic assumptions about business and economics, Muslims have a great deal to contribute when it comes to reviving a transcendent view toward work and service. Yet in highlighting the importance of serving others, we risk the adoption of a different set of misaligned priorities and assumptions.

For many, our renewed emphasis on "work" is quickly misconstrued as an imperative to "follow your passion" or "live your dreams" – a cozy affirmation of our culture's hedonistic refrains about "doing what you love and loving what you do."

For those left unsatisfied by the lure of materialism, it seems like a good replacement. Unfortunately, without the proper arc and aim, it's bound to lead to the same dead ends of self-focus and self-indulgence. Writers like Frederick Buechner further this misconception, defining vocation as melting pot of one's happiness with the humanity's deeper needs.

While vocation can surely manifest in this way, would this, truly, be what it's fundamentally about? If vocation is fundamentally about personal happiness in work, it's "a luxury only afforded to the most privileged on the planet which is bound to lead to a dissatisfaction for those who are doing 'just a job'."

For others, vocation (properly understood) may refer to *love and service to neighbor,* and if we hope gain and absorb that understanding, we'd do well to start with remembering that our "jobs" are only one facet of our vocations and probably not the most important: we also have callings in the family and the society. Our vocations are not just where we find our fulfillment but also where we follow the Qur'an.

God calls every believer to work in their vocations which leads us to serve others. Yet, we also have various other callings that flow from this. Each person should live as a believer in whatever situation God has assigned to them and called them. This situation that God has assigned encompasses the various areas that fall under the rubric of vocation.

Here are five things we should know about vocation:

Vocation is about love and service: As vocation aims to act as a means to serve one's neighbor, the criterion for how to realize each and every vocation should be to ask the question of how one's individual calling would serve his neighbor. Vocation is the specific way in which God calls us to live as a Muslim in the world and serve our neighbor.

Vocation is more than your job: We often use the term vocation in reference to our careers or occupation. But while our jobs are a way – maybe even the most significant way – we serve others, the Islamic concept of vocation is more expansive. It includes all the roles in which we are called to serve the mankind.

Vocation is not self-chosen: A vocation is something we are called to by God. It is not something we choose for ourselves. We discover our vocations by considering what resources God has given us for being a servant (i.e., talents, interests, abilities) and the people he has put in our lives (e.g., parents, children).

You have multiple vocations: We are called to serve in various spheres, such as the family, the workplace, mosque, etc. In each of these we have a vocation – sometimes multiple vocations (e.g., being both a parent of a child and the child of a parent).

The primary vocation of a Muslim is to be a Muslim: While most of our vocations are equal before God, one stands apart from all others: our calling to be a follower of our beloved Prophet Mohammed (s.a.w.w). This is the most important vocation we will ever have in this life.

Given the self-centeredness and materialism of economic pursuits, vocation is a great means of countering this materialism by providing a new meaning and direction. It dismantles the self-centeredness of our age, from the materialistic pursuit of personal wealth to the emotional pursuit of self-actualization. So the Kenyan construction worker and the Bangladeshi woman sewing buttons have vocations. Vocation honors labor that the world looks down upon. Yes, being a wife, mother, sister, etc., is just as important as getting paid for a job – indeed, more important, the family responsibilities being the most fundamental. With regard to our various responsibilities – as spouse, parent, citizen, and worker – we are to live out our faith. Such a doctrine of vocation with its radical, neighbor-centered ethic displaces good works from the realm of the merely spiritual into the realm of the material, the social, and the ordinary. When we serve humanity, we do serve God.

Our modernistic and hedonistic sensibilities will surely resist such a framework, arguing, rather ironically, that all this amounts to a different sort of sentimentalism and emotionalism.

But just vocation is not spiritual frosting for materialism, nor it's also an excuse for pleasure-seeking and "following your passion." Vocation is a means for "jihad" serving human neighbors for the glory of God. Ours is a service not of our own design or choosing, and when we orient our lives accordingly, it's far more powerful because of it.

The classical notion of economy (Greek, *oikonomia*; Latin, *oeconomia*) had been associated with the family, and *oikonomia* was literally the "law" or "rule" of the household.

It was for later thinkers to develop and apply this idea to the arena of social organization of families within society, first as political economy and later as economics as distinct from politics. Debates at origins of neo-classical economics struggled to differentiate between activities that counted as "economic" in the newer sense as being done outside of the family and those that were beyond the realm of economics, that is, within the household. The famed sociologist Max Weber would later develop an understanding of the relationship between Protestant moral teachings and vocation that would increasingly understand the latter in economistic terms. As Weber asserted, within the context of the modern economic order, money can be earned legally as long as there is proficiency and virtue in a calling. A secularized understanding of vocation places the person within an "iron cage" of ever more profit-making, which is far different from those who out of love seek new and better ways of serving others and thereby serving God.

As there are increasing pressures of economic change and uncertainty, many have relished in a range of renewed nostalgias, whether recalling the blissful security of post-war industrialism or the rise of the Great Society and the prowess of the administrative state. Meanwhile, economic progress continues at an increasing pace. Indeed, as politicians attempt to prevent or subvert economic change by squabbling over wage minimums, salary caps, trade barriers, and a host of regulatory fixings, entrepreneurs and innovators are trying to accelerate learning to do less with more. As symbolized in the form of job-killing ATM machines, we live in an age where new com-petitive pressures are met with new efficiencies with astonishing speed. For example, Amazon recently showcased plans for the roll-out of its new conve-nience store concept, in which cashiers will be completely replaced by new "AI-powered technology."

No sooner was the concept hailed as "the next major job killer" than we learned about the next iteration: a supermarket-sized store run almost entirely by robots, leaving room for as few as three human employees. According to an inside source quoted in The New York Post, Amazon will continue to "utilize technology to minimize labor."

Upon hearing such news, Luddites of varying persuasions will surely be tempted to instigate the garden-variety protective and coercive measures to slow or halt the pace of innovation. Yet to do so will only delay or exacerbate the inevitable.

Every technological revolution destroys old jobs. The new productivity takes a different form each time, but it ultimately doesn't have to mean fewer jobs overall. It means a change in the way jobs are defined.

At the beginning of the 20th century, mass production did the same thing to shop production that electronic production is doing to mass production now. It eliminated jobs – at first. Mass production could create many identical units at low cost. The ideal policy was thus to make energy and materials cheap and labor more expensive, thereby creating more mass-market consumers using cheap fuels and electricity. After World War II, governments in the industrialized world did just that, raising the cost of labor by supporting labor unions, establishing payroll taxes, and passing minimum-wage laws. Cheap raw materials and energy, in the form of fossil fuels, came from the developing world. Even though businesses chafed at high salaries, they benefited from the increases in productivity and in demand.

Today, it's energy and materials that are too expensive (or will become so if growth resumes strongly), and they need to be reduced to cut costs. Environmental threats reinforce this incentive. Thus, businesses are redesigning products for smaller carbon footprints, fewer materials, and zero waste. Many products are also being turned into services – prerecorded music into streaming, for example.

According to the Frey and Osborne study from the Oxford Martin School, 47% of today's white-collar jobs in both the United Kingdom and the United States might be automated by 2035. A recent World Bank study suggests that 69% of all Indian jobs are vulnerable to automation.

Forecasters disagree over whether the coming wave of robotic automation will usher in a utopia or a wasteland, but none questions a future where automatons increasingly put human beings out of work.

Prognosticators agree that machines will meet an ever-growing share of man's needs and desires, while homo sapiens retreat into forced idleness. The optimists believe this will free humans to devote hone their higher, God-given faculties. Pessimists worry that a significant portion of the human race will lose its economic wherewithal, as fewer jobs are open to people.

Still others assert that the world will never see a future in which shiftless people relax while robots create more products than they can ever afford. According to them, the automation of existing production will do what it has always done. It will create the conditions for the development of new products, for new demands, and for new, different and well-paying jobs to emerge.

The 21st century differs from the previous business revolutions as AI, robotics, 3-D printing, nanotechnology, biotechnology, and the like are disruptive technological changes. While the debate about whether these technologies may bring superabundance or lasting job losses and lower pay may still be ongoing, neither outcome is plausible.

Former world chess champion Garry Kasparov has a new book out called Deep Thinking. Losing to the computer "Deep Blue" gave Kasparov a close encounter with a member of the race of "intelligent machines." Despite this experience, he tends to see promise rather than threat in their growing power. As Garry mentioned in one of his essays for the Wall Street Journal, as machines replace physical labor, there would be more opportunity for human beings to focus on their minds so that more menial aspects of cognition can be taken over toward providing more space for enhanced creativity, curiosity, and joy.

It is instructive to take a backward leap to 1930 and to John Maynard Keynes looking forward to a bountiful future of superabundance in one of his essays "Economic Possibilities for Our Grandchildren" as stated that human-beings should value ends above means or prefer the good to the useful.

An important point that should be taken into account is that here Keynes was talking about himself and people like him rather than the majority of individuals. Most people may not welcome a future of enforced idleness and leisure, however much opportunity it gave for mental contemplation. The latest round of automation and technological upheaval will not result in a land of plenty co-existing with entrenched unemployment. As Keynes' prediction proved to be wide of the mark, so will be Kasparov's. Free market economics and human nature are the keys to understanding the effect of yet another industrial revolution, since the first began in the second half of the 18th century.

Instead of rendering giant companies obsolete, the current system might complement them by opening other opportunities for wealth creation of another sort. As John Maynard Keynes correctly asserted, someone needs to create demand before innovation and investment can come forward. The last time it was by building houses on suburban land. Yet, how do you create demand now?

That's where emerging economies are important. The so-called developing countries were not included in the mass production surge of the 20th century, because the advanced world was more interested in their natural resources than in their consumer market. Yet, that is changing now. As countries like China and India continue to grow rapidly, they provide demand needed by business producers, including food and materials producers in other emerging economies. These new producers will take advantage of much greater global demand to fund their development, which in turn should increase global demand for capital and consumer goods. It's a new positive-sum game waiting to be set up.

The critical question is: Can a positive-sum game be established among all the world's nations? The need for full global development today is enormous, if only because of the growth in consumer demand that's needed. Even if ISIS is defeated, you must establish enough jobs in the less-developed economies to bring back hope to their populations.

It's essential for every economy to specialize, so that it can participate competitively in global markets. Yet, each piece of territory has to abandon the race to the bottom and define its identity, connected to its history or to strengths that it creates. Its businesses, universities, regulatory priorities, and tax regime must all favor the chosen direction for success, preferably defined by a consensus-building process. The advanced industrial countries may end up specializing in capital-intensive goods and high-level engineering while the lagging countries will have to build their own manufacturing bases. Some may specialize in raw materials-based industries while others may have their own entrepreneurs, innovating in products and services that reflect their culture and identity. Diversity is in the nature of information technology just as much as homogeneity was natural to mass production.

The Industrial Revolution was a gradual process of development comprised of the individual actions of thousands of innovators across time. The dramatic changes in the world have come about partially due to the technological growth, some of which developed out of this revolution of industry. It is not the result of a few "great, singular men," but of many interconnected individual innovations.

Technological growth comes about from the trade and exchange of concepts and ideas. Each innovation builds on previous innovations, connected in a web of technology that encompasses the world (See Appendix A for a complete review of innovation and competency policies). The more and the freer the trade, the more and better technological growth. As trade grows, so does the spread of knowledge and ideas, and the resulting innovation.

Most automation of jobs is only partial, not complete. Even partial automation can lead to jobs losses, of course. But as a whole, automation tends to merely shift the need for human labor from routine, low-skill tasks to more creative, high-skilled functions. Automation leads to fewer elevator operators but more elevator designers, engineers, and repairmen.

This shift ought to be lauded by Muslims. While we should rightly be concerned about the employment prospects of low-skilled workers, we should not become nostalgic for the mind-numbing, back-breaking work that automation has made obsolete. Too often we treat "jobs" as if they were an inherent good (at least if they pay a "living wage.") But not all jobs are created equal. Some jobs that may benefit our neighbors' bank account may also be crushing their soul.

The rapid adoption of computerized automation has the potential to increase job satisfaction for entire occupations that have previously been dangerous, dirty, and demoralizing. In looking at the future of work, we therefore must look not only at the wages that a job will pay but also at the price such work requires of our neighbors. We can let the robots take over the parts that a machine can do so that we may use our God-given human abilities for more ennobling tasks. This can be achieved by improving upon our intuition and creativity while working hard. It is sad that today's millennials – many members of my own generation including myself – have a hard time in accepting this simple reality.

As Muslims, especially, we should remember that adapting to economic change is fundamentally about adapting to *human needs* and aligning the cultivation of our minds and the toil of our hands with love for and service to our neighbors. Rather than nostalgic pining for ages and economies past, we can move forward – not by manipulation via the levers of government power, but by building on our God-given intuition and creativity, staying pro-active in our response to the shifts that are sure to come.

When the economic conditions change, the voice of God will speak, wisdom will come, and we can move forward energetically and creatively, leaning not on our own understanding. We may think that certain forms of such destruction signal our end, but when we align our hands to anticipate the dynamism of a new set of needs, the ultimate solution may surprise us after all.

Leaders would have to understand their role in this crucial moment, move to open a consensus-building process, and be determined to take bold measures. Their efforts, hopefully supported by business and society, could be the basis for the global golden age of the information economy.

According to the execute chairman of the World Economic Forum, Klaus Schwab, the following revolutions occurred throughout the history of world economy:

- Steam power mechanizing production;
- Electric power facilitating mass production;
- Electronics and information technology automating production.

This latest industrial revolution might possibly turn out to be more profound and fast-moving than earlier revolutions. Accordingly, it might turn out to be more disruptive and cause more transitory unemployment. However, there is no basis for believing that it will result in permanently higher levels of unemployment, nor in enduringly depressed incomes.

Recent historical experience is instructive. Workforce participation has risen since World War II has finished, despite increasing automation. In the United States, the participation rate was less than 60% in the 1950s compared with well in excess of 60% in more recent years. Participation has similarly risen in, say, the United Kingdom and in Australia since the 1970s. It is true that underneath these numbers is rising participation by women against falling full-time participation by men. Yet, the plain fact is that overall workforce participation has risen rather than fallen, as have real hourly wages, over the past 50–60 years of profound labor-saving technological changes.

There have been a large number of technological changes since the first Industrial Revolution began in Britain around 250 years ago. If these had cumulatively dampened employment, very few people would now be employed. Moreover, it is the new norm that families with children require two incomes in order to pay the bills and live comfortably.

It is certainly possible that the latest industrial revolution will produce more dislocation than experienced in the past. But it is essential to take account of unrequited human wants and the symbiotic relationship between those making products and those buying them.

There are more than seven billion people in the world, very few of whom have all that they want. We have a system of capitalism that generates a balance between supply and demand – and which critically depends on sufficient numbers of people being employed and being paid enough, to purchase goods or services. The system simply will not work for any length of time if there is an imbalance between automated production churning out vast quantities of products and a growing army of unemployed who can't afford them. It simply can't and doesn't work that way. Temporary imbalances can exist and produce recessions. But, left alone, these are part of a process of the economic system moving back into alignment.

Long-term mass unemployment is not a technological phenomenon, neither is it an economic one. It is a political one. The benefits come in the form of a range of different jobs and of increased supply and variety of goods and services – for the poor, the rich, and the not-so rich. In the latest industrial revolution, there is nothing to fear but fear itself – and inept and interfering governments.

Given the breakneck pace of improvements in automation and AI, fears about job loss are taking more space in the cultural imagination. Symbolized by recent concerns about Amazon's "job-killing" grocery-store robo-clerks, the anxiety is palpable and persistent.

Using elaborate models and forecasts to affirm such fears, economic planners and doomsayers predict the rise of robot overlords and the demise of human labor. Such estimates paint a dismal economic future wherein humans are pushed to the side with little to contribute and even less to gain. Yet, could it be that this picture might be missing something?

As Ross Gittins explains, the common modeling which has been suggested by scholars such as Frey and Osborne includes significant errors, oversights, and inconsistencies when applied to the real world. To give a more specific example, Frey and Osborn asserted that surveyors, accountants, tax agents, and marketing specialists were automatable occupations, whereas Australian employment in these has grown strongly in the past 5 years. According to Frey and Osborne, the need for dexterous fingers is an impediment to automation, yet their method predicts that there is an automation probability of 98% for watch repairers.

Furthermore, Frey and Osborne's modeling assumes that if an occupation is automated then all jobs in that occupation are to be destroyed. So the advent of driverless vehicles is assumed to eliminate all taxi drivers, truck drivers, couriers, and more.

Moreover, these scholars make the assumption that if it's technically feasible to automate a job, there would be no need for employers to decide due to reasons of profitability. Similarly, these scholars assert that there will be no shortage of the skilled workers required to set up and use the automated technology.

Apart from their predictions about high-level trends – that certain jobs, sectors, and industries will indeed be largely automated – these scholars fail to recognize or account for the unseen and unforeseen developments that result from automation. As a result, the transformative role of human potential and ingenuity is ignored amid technological progress. Such a modeling involves no attempt to take account of the jobs created, directly and indirectly, by the process of automation. No one gets a job selling, installing, or servicing all the new robots. Competition between the newly robotized firms doesn't oblige them to lower their prices, meaning that their customers don't have more to spend – and hence create jobs – in other parts of the economy. All that happens, apparently, is that employment collapses and profits soar. But if it happens like that it will be the first time in 200 years of mechanization and 40 years of computerization.

Such an outlook requires not only a static view of the economy, but also a remarkably dim view of human creativity and possibility. If we look to history, we see that automation has led to greater prosperity and productivity, making more room for humans, not less.

This is precisely because we are not mere machines, consigned to junk yards when particular solutions or services are rendered obsolete. Rather, we are creative and imaginative human persons created as stewards of the only and one Creator. We are fully capable of adapting, mobilizing, and innovating our modes of service to be in line with His purposes in the earth. When the economic conditions change and mechanization or automation replaces old ways of meeting human needs, innovation comes and new human services are created.

Automation will continue to disrupt our old ways of doing things. But knowing what we do about the past and the future of human possibility, we needn't be fearful of our own position and power. As we survey the barrage of predictable reports about the end of human labor or the rise of robot dominance, let's be sure to wield our hope and skepticism accordingly.

Moreover, we'd do well to remember that doing less with more also means having more time and resources for more. Data continue to demonstrate that technology creates more jobs than it destroys, and we'd do well to seize on that reality with optimism and a thirst for the next opportunity for creative service.

That optimism requires more than just a change in attitude or orientation if we hope for the fruits to endure. As researchers like economist Tyler Cowen have explained, our vocational priorities need a drastic rehaul if we hope for our skills and study to apply in the economic order. As Cowen asserts, with the increasing automation and productivity levels of machines as well, the question of where human beings should be located within this picture should be reconsidered. This means that questions such as what would be the criteria to make progress in this new digital economy should be given serious thought.

As the authors Brynjolfsson and McAfee's quote in their book "Race Against the Machine" from the economist Clark, the population of horses to be used for work in England vanished later on as the workers were replaced in the 19th century with the arrival of the combustion engine and thus that the low wage did not pay for the feed of the horses. These authors assert that unarguably technology enables mankind to achieve progress in an unparalleled way. Due to the ICT profound structural changes are taking place. To give a specific example, similar to the jobs being eliminated due to steam power and the internal combustion engine, several jobs that we hold today may be eliminated given the connectivity and automation opportunities provided by our digital age.

As the Industrial Age came in two main stages, namely steam and rail-roads followed by electricity and the engine, the ICT revolution started with the punch card and telephone lines not to mention today's smartphones which may be just the beginning of what this age will bring.

On the other hand, given this increasing level of connectedness and complexity provided by the ICT revolution, it becomes more difficult to discover how we can provide virtue and value in our apart from earning merely a compensation to pay our bills. Each job in fact has the potential to provide an opportunity to collaborate with God, often in ways that we may never see.

If individuals were asked about their career decisions, most of them would admit that they have been influenced by chance events to a great extent. So these seemingly random perturbations in addition to the initial starting point disrupt the ultimate trajectory of personal careers. Among the most common "random perturbations" are:

Limited options: There is a good chance that one's future job title will be one which has never been imagined before as they even did not exist once. Yet, despite the lack of complete control on the trajectory of our careers, we can still influence the initial conditions rather than choose a path based on restricted conditions.

Miswanting: Several research studies suggest that we are terrible at pre-dicting what will make us happy in the future. This phenomenon has been referred to as "miswanting." Even though we may assume that money and status may provide pleasure, in the longer run we may discover that we should be seeking another mix of aims as our desires and motives will change radically over the next 10 years. In other words, the dreams we thought our future self would follow were not really what we wanted at all.

Focus on skill sets: The skill clusters we obtain, rather than our prefer-ences, have often specified what job will become available to us. As these skill sets have a significant impact on the career path, we should choose them carefully.

Who knows you is more important than a resume: A better version of the cliché "It's not what you know but who you know" should be "It's not what you know but *who knows you*." As we try to expand our network, our career will (mostly) take care of itself.

Lack of insight on where the journey will take: Although one's personal circumstances may not be particularly promising, they will change eventually.

More importantly, as Muslims we should bear in mind that we can be anything that *God* wants us to become. Our Creator clearly sees our path even if, from our limited perspective, it may look like a chaos. If we keep seeking his guidance by reading the Qur'an and obeying his commands, eventually He will enable us to practice our skills and creativity provided by Him into useful service. By trusting God, we can enjoy the exciting journey ahead of you.

Technology has played an important role in lessening the existing supply of labor while keeping many potential workers minimally satisfied and happily engaged in the comfort of their parents' homes.

This applies to consumption-oriented technology in general, but with video games in particular, we see a unique form of entertainment that often looks and feels like the best parts of the very workplace we're seeking to avoid. Unlike TV or movies or social media, video games move beyond idle diversion, creating illusions of toil and, in turn, earned success.

As Peter Suderman explains, in recent years, video games have evolved into something strikingly similar to what some might call "meaningful work". Video games can be considered as a kind of employment simulators as they include basically a series of quests comprised of mundane and repetitive tasks. So it is perhaps not surprising that video games are increasingly replacing work for many young men.

Video games serve as a buffer between the individual and despair as even the most open-ended games offer a sense of progress and commitment. Yet, although video games may indeed provide an initial rush of merriment or serve as a reasonable method for emotional "coping," the bigger question is whether that's actually where it stops. At what point does our cultural obsession with technology, and video games in particular, move from innocent leisurely consumption to a culture-wide replacement of meaningful production?

Whatever benefits these games may provide in the short term, and whatever emotions their "simulations" may satisfy, we should be careful to make proper distinctions about the fundamental aim, which impacts all else.

We should remember that as Muslims, meaningful work is about much more than simply overcoming obstacles, achieving "organizational success," or becoming an expert in a particular skill or industry. Through means of work, we serve not only others, but also God himself and He, in return, weaves the work of others into a culture of service that makes our work more rewarding. This is why work gives meaning to your life and to mine.

This is the basic difference between work and leisure, and it's a distinction that's bound to shape our attitudes and imaginations as we prioritize the hours of the day. Play can be considered as a kind of recreation, as it prepares the individual to do his work better. Play is fun and relaxing, because it is always an end in itself. The desire that leads to playing is satisfied in the doing.

To be clear, leisure can be a very good thing. It plays an important role and can bring its own form of meaning in refreshing the human mind and spirit. Yet, when we confuse it for something else – seeking meaning in the feats and victories themselves – we move closer to embracing the counterfeit as an idol and calling it something else.

As long as an effort ends in the satisfaction of the self, it cannot be called as work. Regardless of the type of work undertaken, what the self gets engaged for the sake of its own satisfaction rather than realizing God's commands for serving mankind, this will not carry over into Afterlife, and hence will become a burden for the soul itself.

Given the impact of these distractions on the individual lives of workers and the economy in general we should keep remembering that while we work, we serve others unto the glory of God. In play, we play.

We should be excited about this future. Vast new opportunities are coming that will alter the course of human history. Technological change today is already occurring faster and more substantively than we realize. As with any change, however, there will be winners and losers.

Changes are happening so fast that as jobs and career opportunities are being destroyed, the ability to retrain or reset expectations isn't keeping pace.

As Muslims, especially, we should remember that adapting to economic change is fundamentally about adapting to human needs and aligning the cultivation of our minds and the toil of our hands with love for and service to our neighbors. Rather than nostalgic pining for ages and economies past, we can move forward – not by manipulation via the levers of government power, but by building on our God-given intuition and creativity, staying pro-active in our response to the shifts that are sure to come. As AI and the subsequent technology continue to improve, we needn't be fearful of our own position and power. We are not mere machines, but creative and imaginative human persons fully capable of adapting, mobilizing, innovating our modes of service to be in line with his love and purposes.

When the economic conditions change, the voice of God will speak, wisdom will come, and we can move forward energetically and creatively, leaning not on our own understanding. We may think that certain forms of such destruction signal our end, but when we align our hands to anticipate the dynamism of a new set of needs, the ultimate solution may surprise us after all.

6

Trade Dimension of Technology

Trade is deeply grounded in a proper understanding of our human nature. In the process, all parties hone their unique skills, develop bonds of mutual affection, and prosper together. Trade should not be spoken of using the vocabulary of war, of "win–lose." Nonetheless, when discussing trade, economists and politicians use the terms of "we" vs. "them" – Americans vs. Chinese, British vs. European, and so on.

This is where the wisdom of Islamic social teaching may strengthen the case for the freedom to trade. Of course, it is not just trade in raw materials that God has enabled for us to engage in. Each human being is unique, with his or her own talents and capacities, so trade allows us to exchange the diverse fruits of our labor. By restricting trade, politicians undermine, not only trade as such, but also the specialization and realization of human talents. Naturally, it is not "America" or "Europe" that trades with "China," but American and European individuals, families, communities, and companies that trade with their Chinese counterparts. When viewed this way, it becomes apparent just how meaningless the obsession with national balances of trade is, which assume that it is countries that trade, rather than the individuals living there.

When rich economies surround themselves with unscalable trade barriers, it is not only the consumers of the rich countries who suffer, but also the producers of the poor countries. The paramount example of this is agriculture, whereby the leaders of the rich states protect their (already rich) farmers, inhibiting the development of farmers in the poorer parts of the world.

As freedom and opportunity are reaching new corners of the world, technological innovation is allowing us to do more with less. Yet in America, we are no longer living in the safe, secure, insulated, post-war era. It's the same old story of creative destruction but at a new, break-neck speed: more global, more rapid, more dynamic. Through this lens, many of the primary drivers of our newfound prosperity – innovation, automation, offshoring, and trade – are also the drivers of our disruption.

So what is the Muslim's response to such disruption? Is there a way of viewing these constant "threats" to our jobs, comfortability, and convenience as opportunities?

As Muslims, many of us appreciate our "spiritual tools" for such seasons – prayer, fasting, worship, gratitude – and each of these is important. Yet, if we misunderstand God's design and purpose for business and economics, Muslims are at risk of misapplying these same tools, responding out of a mindset of security and scarcity rather than risk and abundance.

Work is first and foremost, service to others, and thus to God – or service to God, and thus to others. From the garbage man to the school teacher to the doctor to the microchip engineer to the software developer to the father and mother, all of our work is about service to our neighbor.

When we shift our perspective toward God and neighbor, everything flips upside down. Work is no longer about "following your passion" or self-actualization, though that may be a byproduct. It's about obedience to God. Work and career are no longer about personal provision, though that will likely be a result. They're about providing for others. Work is no longer about protecting our turf or sitting still in our "niche. It's about creativity, inclusion, collaboration, and competitive development. From here—and only from here—can we effectively apply the range of spiritual tools God has given us bringing prayer and prophecy, wisdom, and discernment, Islamic transformation to all areas of our work, from the assembly line to the board room to the Silicon Valley garage to the home nursery.

When economic change hits, that fundamental switch makes all the difference, turning signals of disruption into signals for creative service.

The temptation to dwell on the illusion of economic security will remain strong – to cherish and fight for the comfortable control we've enjoyed thus far. Yet, to do so requires us not only to succumb to an unworkable fantasy about the global economy, but to distort God's design for work: to give way to selfish impulses, to suppress our own creative potential, and to exclude the creativity of countless others.

No country insulated from its competitors, whether we pretend to be or not. That is not cause for fear and territorialism and protectionism. Rather, it is a good and beautiful and promising thing, if only we'd respond accordingly – reorienting our hearts and hands from a work that secures, consumes, and collects to one that serves, creates, and sustains.

Certainly, people can work just because they want a paycheck to spend on themselves alone. That might be greedy, but we need to be careful not to confuse profit with greed. People work in order to profit, but profit is not good or evil in itself. That judgment depends on the circumstances in which it was gained and the use to which it is put. Our work itself is service to others. If it wasn't, they wouldn't pay us to do it in the first place, and most people wouldn't want to do it for free. It's an exchange. The division of labor is the phenomenon that the more the manufacturing of individual components of an eventual finished product can be broken down into separate jobs, the more efficiently it can be produced. Adam Smith offered the classic example of the pin-maker in which 10 people working alone might be able to produce 10 pins total in a single day, maybe up to 200 (20 each) if they were really good, but nowhere near 48,000. When people work together, they are able to multiply the fruits of their labors far beyond what they could each do alone.

Another way to think of it is the power of human cooperation. The division of labor implies "teamwork." God made us to flourish in collaboration with each other.

It is worth noting as well that this mass production did not in any way change the quality of the pins produced. Sometimes that is the case with products today, but it is not necessarily true. In this case, it was simply by splitting up the labor required to make a pin into each of its parts and then assigning a single task to each person that made all the difference. Because they could make so many so much faster, they could lower the price to consumers while still making astronomically higher profits. It's a win–win.

Our world provides plenty of examples for this. Consider the production of a book. If you are reading the print version, the paper came from trees that were felled by lumberjacks, made into paper in factories, and then shipped to a printer. Similarly, the ink for the words and the cover had to be manufactured, too. All of the factories involved used tools that had to be made somewhere else, by someone else, at some time before. All of the vehicles used to transport the capital that would become this book had to be made by people all over the world, working to provide for their families and, unknowingly, to provide this book for you. If you're using an e-reader, well, there are far more people and resources involved.

Nearly every product, every fruit of the cultivation of creation, connects us with nearly every other human being on the planet. Their collective contributions make this book more affordable while also benefiting more people in the process. In this way, our work connects us with other people, serves their needs through products and property, provides for us, and fulfills one of the purposes for which God made us.

On the other hand, can we think that there is a connection between economics and belief?

The overlapping influence and impact of distinct cultural spheres – what anthropologists call the "functional integration of culture" should not be disregarded when talking about faith and economics.

According to anthropologist Darrell Whiteman, every culture can be understood as having the following three interconnecting sectors:

(1) An economics and technology sector,
(2) A social relationships sector, and
(3) An ideology and belief sector.

These sectors are so integrated that if you create a change in one of them, it automatically influences the others, to create a change there.

Nobody knew what the next epidemic or natural disaster may be, yet God has already been working to provide a solution through some non-profits or humanitarian relief efforts or places of worship such as mosques at the mundane levels of economics and technology.

The response would be a revival. While the ball may end rolling up in an ideology and belief shift, it is also related to the economics and technology sector. In the end, individuals will realize how they socially interrelate with these Islamic associations and mosques in different ways, such that their ideology and belief was changed.

For Muslims, our focus has to remain on the bigger picture – economic, social, religious, and otherwise. Whatever victories we achieve are where active and attentive "executive stewardship" actually begins.

Silicon Valley certainly has a reputation for innovation and risk. But Islam? Businesses designed not only to innovate but to pursuing business as an "intimate" adventure with God? That seems unlikely.

Every entrepreneur would acknowledge the high failure rate for startups, yet creativity and boldness are part of the "package" of business. We should all try to figure out ways to harness human creativity in a way that will better lives, and acknowledge that our faith feeds this.

In a free economy, we each serve distinct roles as both producers and consumers. As producers, we create and serve, leveraging the work of our hands to meet the needs of our neighbors. As consumers, however, we look to ourselves and our own needs. Consumption is a good and necessary thing, but it introduces a range of unique pitfalls and temptations. "Consumption is obviously necessary – there would be no economy without consumers," explains Raymond J. de Souza. Yet, we distort the very value and purpose of economic freedom. Consumerism distorts our fundamental notions of identity and personhood, yielding to the severe moral confusion we see today. In our digital era, the poisonous effects of consumption extend far beyond the mere exultation of material things. Using various social networks, the individual often becomes – in his own mind or in the view of others – primarily an object of a network who consumes for attention rather than a subject who does goods in order to give himself to others. Whereas in capitalism, we face a constant temptation to reduce ourselves to consumers – mere "objects of material things"; in the digital age, we face a constant temptation to reduce ourselves to consumers of digital media – mere "objects of material things."

Consumption eventually brings the dismal features of economic pressure and economic warfare to various levels of culture, yielding ripple effects that predictably pollute the waters of an otherwise free society. Alas, what we begin to see in the digital sphere – spreading and manifesting among free, diverse peoples – begins to look ever more similar to government-induced conformity, inspired not by spirit or conscience or creativity, but by force and intimidation.

Freedom is part of human essence. In an age that's grown servile to the idols of consumption, we have the opportunity to show what true freedom looks like, showing through the work of our minds, the stewardship of our resources, and a tolerance toward freedom of conscience, that all economic activity – even consumption – is destined for the service of the mankind.

Only God is all-knowing. We are finite and limited, so prices help us disperse knowledge efficiently. And, just as there are built-in interdependencies in other parts of creation – and particularly with respect to using the gifts each of us has been given – the knowledge problem also reveals our inherent interdependence. Prices communicate information we need to streamline those interdependencies and serve each other through trade.

Ultimately, overcoming the knowledge problem to make good decisions with what we've been given is good stewardship and is part of God's design and desire for his creation.

7

Dark Dimension of Technology

"Terrorists use the Internet just like everybody else"
–Richard Clarke (2004)

7.1 Introduction

Conversations around surveillance focus mostly on the use of encryption in communications technologies. This debate has caught increasing attention due to the decisions of Apple, Google, and other major providers of communications services and products to facilitate end-to-end encryption in certain applications, on smartphone operating systems, while terrorist groups try to use encryption to hide their communication from surveillance.

Most recently, terrorist attacks in San Bernardino and Paris along with increasing concern about the terrorist group ISIS have shifted the focus onto the issues of surveillance and encryption. These developments have led to new invitations for the government and private sector to work together on the going dark issue.

According to Resnick (1999), there are types of Internet politics. Resnick defines "Politics within the Net" as the political life of online communities and related activities that impact the life off the Net only to a minimal extent. Another category that Resnick (1999) defines is "Politics which Impacts the Net" that involves policy issues emerging due to the use of the Internet as both a means of mass communication and a tool for business. The last category that Resnick (1999) defines is "Political Uses of the Net," which involves the use of the Internet by citizens, activists, or government to fulfill political goals that don't relate to the Internet *per se*. In other words, this relates to influencing political activities offline (1999, pp. 55–56).

This section mainly focuses on the third category,' especially terrorist' use(s) of the Internet.

Rash (1997), referred to eight different uses of the Net by political groups and categorized them as follows:

- Tactical communications,
- Organization,
- Recruitment,
- Fundraising,
- Strategic positioning,
- Media relations,
- Affinity connections, and
- International connections (1997, pp. 176–177).

Similarly, in terms of the main uses of the Net by terrorists, Furnell and Warren (1999) made the following classification: propaganda/publicity, fundraising, information dissemination, and secure communications (1999, pp. 30–32).

While Cohen (2002) provides his readers with a similar list of uses (pp. 18–19), Thomas (2003) explains the terrorist uses of the Net in more detail. In addition to his discussion of several possible terrorist' uses of the Internet, Thomas (2003) refers to "cyberplanning" as the digital coordination of a comprehensive plan across geographical boundaries that may or may not result in bloodshed (p. 113).

Furthermore, Weimann (2004) refers in one of his reports for the US Institute of Peace to the following ways of the terrorists' use of the Internet: psychological warfare, publicity and propaganda, data mining, fundraising, recruitment and mobilization, networking, information sharing, and planning and coordination (pp. 5–11).

Given this overlap among the uses of the Net by terrorist, Table 7.1 analyzes and explains in more detail these classifications.

The vents with regard to the introduction of built-in encryption by Apple, Google, and other technology companies can be summarized as follows:

- In September 2014, Apple announced its decision to use a default encryption of the password-protected contents of its devices in the then-next version of its mobile operating system, iOS 8. Data generated by many of the system applications on iOS 8 and later versions are encrypted when data is stored locally on the phone, in transit, and stored on Apple's servers. The decryption keys are tied to the device password and only stored locally on the phone.

Table 7.1   Terrorist uses of the net

Author(s)	Furnell and Warren (1999)	Cohen (2002)	Thomas (2003)	Weimann (2004)
Uses	(1) Propaganda and publicity (2) Fundraising (3) Information Dissemination (4) Secure communications	(1) Planning (2) Finance (3) Coordination and operations (4) Political action (5) Propaganda	(1) Profiling (2) Propaganda (3) Anonymous/ Covert communication (4) Generating "Cyberfear" (5) Finance (6) Command and control (7) Mobilization and recruitment (8) Information gathering (9) Mitigation of risk (10) Theft/Manipulation of data (11) Offensive use (12) Misinformation	(1) Psychological warfare (2) Publicity and propaganda (3) Data mining (4) Fundraising (5) Recruitment and Mobilization (6) Networking (7) Sharing information (8) Planning and coordination

- Afterward, Google announced that Lollipop, its next version of Android OS, would enable device encryption by default.
- Then, WhatsApp, the popular instant messaging service for smartphones now owned by Facebook, announced its support for TextSecure, which is an end-to-end encryption protocol.
- In March 2015, Yahoo introduced a source code for an extension that encrypts messages in Yahoo Mail, though it requires users to run a key exchange server.

All of these events bring to the mobile world some of the available technologies that were not enabled by default for personal computing operating systems, such as Apple's FileVault and Microsoft's Bitlocker.

The most significant aspects of these announcements are that the encryption takes place using keys solely in the possession of the respective device holders which is enabled by default.

When encryption is seamlessly integrated into a communication software, a user does not have to take any actions to encrypt or decrypt messages, and much of the process occurs on the back end of the software. In fact, an average user might not be able to recognize the difference between an encrypted and an unencrypted message. When these options are enabled by default on popular devices and platforms, like the iPhone, a large part of communications is encrypted.

It should be kept in mind that this is not the first debate about the public's ability to use encryption and the government's ability to access communications. Before the PC and Internet era, telecommunications companies were providing a means for government actors to wiretap voice communications – in particular on the legacy telephone system. Within the framework of the CALEA (Communications Assistance to Law Enforcement Act), US telephone companies would ensure that their networks could be wiretapped, with appropriate legal process, as network technologies moved from analog to digital. Later on referred to as the "crypto wars," government access to encrypted communications has become a hot debate and restrictive policy as the government ultimately was relaxing many export-control restrictions on software containing strong cryptographic algorithms in 2000.

Later on in October 2014, shortly after the announcements from Apple and Google, FBI Director J. Comey stated that as the law hasn't kept pace with technology, an important public safety problem arose. He referred to the issue as "Going Dark," and mentioned that those responsible for protecting citizens cannot always access the evidence which is required to prevent terrorism despite their legal authority because of their lack of technical ability.

According to these statements, the going dark problem is exasperated by the stronger encryption standards and advent of default encryption settings on both networks and devices. As a result, the use of encryption may constrain the ability of law enforcement and the intelligence community to investigate and prevent terrorist attacks.

Although much of the debate was about whether the issue is for companies like Google and Apple to preserve access to user data, the FBI or other stakeholders of the law enforcement and intelligence communities made no formal proposals. There has been only an invitation for collaboration among academics, Congress, privacy groups, and others in order to be able to address the needs of various competing legitimate concerns. Specifically, the private sector was asked for help in identifying solutions that provide the public with security without disappointing lawful surveillance efforts (See Appendix B for a detailed discussion on cyber security).

Other countries in the EU have been witnessing similar debates. In the United Kingdom, an outright ban was proposed by the Prime Minister on end-to-end encryption technologies following the January 2015 attacks at the *Charlie Hebdo* offices in Paris. Following the more recent November attacks in Paris, French authorities also started to question policies about the availability of encryption software.

Many geopolitical partners to the United States are actively engaged in discussions about promoting cyber security and the appropriate limits of surveillance across borders. For example, due to the concerns about the US intelligence community's ability to access data, the US-EU Data Protection safe harbor, which provided a legal framework for commercial cross-border data flows, was recently ruled invalid by the Court of Justice of the European Union. The U.N. also preferred a limited extent on encryption because of seeing it as a required aspect for the exercise of the right to freedom of expression (Emery et al., 2004).

Meanwhile, many US companies must also answer to governments of foreign countries as they have to make a difficult decision due to the pressure from foreign government agencies to produce data about citizens abroad. Several companies refuse to change the architecture of their services to allow such surveillance. However, if the US government were to require architectural changes, surveillance would be made easier for both the US government and foreign governments. Needless to say, the well-developed legal doctrines, and redress mechanisms as part of the US government's surveillance activities are not available worldwide.

Although use of encryption may hinder surveillance, it may not be impermeable. For example, intrusions at the end points, a technique often used in law enforcement investigations may not be prevented by encryption. Also metadata, such as e-mail addresses and mobile-device location information that must remain in plaintext to serve a functional purpose, is not protected by encryption. Through means of cloud backups and syncing across multiple devices, data can also be leaked into unencrypted media (Emery et al., 2004).

The "going dark" metaphor refers to the state of communications becoming out of reach and making us blind as an aperture is closed. The recent trajectory of technological development is not captured by this metaphor.

The following trends underpinning government access should be taken into account (Emery et al., 2004):

- Several business models rely on access to user data.
- Products are increasingly being offered as services, and centralized architectures are more common due to cloud computing and data centers. A service that involves a relationship between vendor and user lends itself much more to monitoring and control in comparison to a product,

where a technology is purchased once and then used without further vendor interaction.

- The Internet of Things (IoT) offers a new frontier for networking objects, machines, and environments in new ways. For example, when a television has a microphone and a network connection, and is reprogrammable by its vendor, it could be used to listen in to one side of a telephone conversation taking place in its room – regardless of how encrypted the telephone service itself might be. These aspects point to a trajectory toward a future with more opportunities for surveillance.

It should not be inferred that the problem will necessarily be solved by making data available or by providing government with the ability to gain access. Rather, the forces opening new opportunities for government surveillance mean that, whatever the situation with iOS 8 encryption versus its predecessor, "going dark" does not clearly refer to the long-term landscape for government surveillance (Emery et al., 2004). Any debate about surveillance capabilities today with potential lasting policy impacts should consider these larger trends (Hoo et al, 1997).

7.2 Encryption Runs Counter to the Business Interests of Many Companies

Implementation of end-to-end encryption is discouraged by current business models and therefore, there is an impediment to company and government access.

For the past 15 years, advertising has been the dominant business model of consumer-facing Internet companies that used ads to subsidize free content and services. Recently, due to the shift toward data-driven advertising, the related technology relies on user data for targeting ads based on demographics and behaviors. Companies such as Google and Facebook try to make behavioral assessments to match ads to individuals on the fly based on behavioral patterns, search queries, and other signals collected such as location, demographics, interests, and behavior (Emery et al., 2004).

Given the preference of companies to have unencumbered access to user data with relevant privacy settings which restrict dissemination of identifiable customer information, the implementation of an end-to-end encryption would be in conflict with the current advertising model (Zanini, 1999). In order not to get their revenues curtailed in this lucrative market, many Internet companies will continue to fulfill the government's requests of providing access to communications of users.

For those companies who require to offer features in cloud services, end-to-end encryption is not practical as they need to access plain text data (Stern, 1999). Cloud computing enables businesses and individuals to extend their computing resources through the Internet at remote data centers, much like a utility service. For example, Google's various features in its web-based services must access plain text data, including full text search of documents and files stored in the cloud. In order for such features to work, Google must be able to access the plaintext. Also, Apple's encryption does not include all of its services although end-to-end communication in some of its applications is encrypted. This includes the iCloud backup service, which enables users to recover their data from Apple servers. Although Apple does encrypt iCloud backups, it holds the keys so that users who have lost everything are not left without any solution. As Apple holds the keys, it can produce user data that resides in iCloud whenever legal processes request the company to do so.

One of the main reasons why businesses do not make a shift to encryption or other architectures is that encryption schemes often add complexity to the user experience. For example, in the Android ecosystem, wireless providers and handset manufacturers may control smartphones by developing customized versions of the Android operating systems. Due to the additional resources required to make the customized features compatible with newer versions of Android, these companies have little incentive to update older phones. Moreover, although the next version of Android released by Google may contain apps that support end-to-end encryption, a manufacturer may change the software so that it includes its custom apps that do not support encryption (Garrido and Halavais, 2003). If the ecosystem is fragmented due to commercial interests in retaining access to plaintext communications, encryption is less likely to become all encompassing.

7.3 Other Surveillance Mechanisms: The Internet of Things and Networked Sensors

According to analysts and commentators representing the conventional wisdom, the IoT is the next revolution in computing which can shift the way we interact with our surroundings (Garrido and Halavais, 2003).

Several companies such as Amazon, Apple, Google, Microsoft, Tesla, Samsung, and Nike are all developing products with embedded IoT functionality, with sensors ranging from proximity sensors, microphones, speakers, barometers to infrared sensors, fingerprint readers, and radio frequency antennae with the aim of sensing, collecting, storing, and analyzing detailed information about their surrounding environments (Garrido and Halavais, 2003).

In February 2015, it was asserted that Samsung smart televisions were capable of listening to conversations due to an onboard microphone. Samsung warned its users in one of its statements within its privacy policy that they should be aware that any personal or other sensitive information, spoken will be among the data captured and sent to a third party via means of the Voice Recognition. A cloud infrastructure through a network connection was utilized to send the voice data to a remote server for processing and interpretations of that data back to the television as machine-actionable commands.

In a similar vein, by means of an onboard microphone in a laptop or desktop computer Google's Chrome browsing software supports voice commands. In order to activate the feature, the user has to say "OK Google" so that the resource-intensive voice processing occurs on Google's remote servers.

Given this impact of IoT, intelligence agencies may start to request orders from Samsung, Google, or others to push an update. It is crucial to appreciate these trends and to think carefully about how pervasively open to surveillance our surroundings should be.

As the expected substantial growth of networked sensors and the IoT there is potential for the surveillance mechanisms to be drastically changed (Garrido and Halavais, 2003). Through means of the still video, audio, images captured by these devices, real-time intercept and recording with after-the-fact access may be realized (Garrido and Halavais, 2003). So the inability to keep track of an encrypted channel could be offset by the ability to monitor from afar an individual via another channel.

Also, it should be kept in mind that metadata such as location data from cell phones, telephone calling records, header information in e-mail will mostly remain unencrypted in order for various systems to be able to operate among each other. This entails crucial surveillance data that was unavailable before these systems became widespread (Garrido and Halavais, 2003).

Governments and related stakeholders should address threats to fundamental human rights such as privacy and personal data protection by acting both within their own jurisdiction and in cooperation.

- Interception of communications, data collection over the Internet, and its analysis by law enforcement agencies and other related stakeholders must address clear objectives that are in alignment with the principles of necessity and proportionality (Garrido and Halavais, 2003). Goals such as gain of political advantage or exercise of repression are not legitimate.

- In case of abuses, an appropriate redress of law should be possible and individuals whose right to privacy has been violated due to arbitrary surveillance should be offered access to an effective remedy.
- Users of free or paid services on the Internet must be equipped with an understanding or a choice over the deployment range of their data with regard to its commercial use, without being excluded from the use of these services (Garrido and Halavais, 2003). In the case of a security breach, a redress should also be offered by these businesses.
- Governments should not ask third parties to build "back doors" to access data as this might weaken Internet security. Integration of privacy-enhancing solutions including end-to-end encryption of data in transit and at rest should be encouraged.
- In collaboration with business sector and the civil society, governments must provide public training on cyber security to foster development of more secure and stable networks globally (See Appendix B for a detailed discussion on this topic).
- Due to the trans-border nature of digital intrusion, the ability of the target government to investigate and prosecute the related parties with regard to that intrusion is curtailed. There must be mutual support among states in order to prevent damage and to deter future attacks.

The debate over encryption which poses difficult questions about security and privacy must be approached from different perspectives.

From the perspective of national security, the question this debate poses is whether access to encrypted communications would also increase our vulnerability to digital espionage and other threats and also whether nations that do not embrace the rule of law would be able to exploit the same access.

From a civil liberties perspective, the question this debate poses is whether preventing the government from getting access to communications under circumstances that meet regulatory requirements will establish the right balance between privacy and security, particularly when terrorists and criminals seek to use encryption to evade government surveillance.

These questions should be examined by focusing on the trajectory of surveillance and technology. As the enhanced availability of encryption technologies hinders government surveillance under certain circumstances, the government is losing some surveillance opportunities. Yet, the combination of technological developments and market forces will provide the government with new opportunities to collect critical information from surveillance.

Given the prevalence of network sensors and the IoT, new questions about privacy over the long term might be raised. So both the responsibilities of technology developers and operational procedures and rules should be re-considered to support the law enforcement and intelligence communities navigate the related issues.

8

Final Remarks

Is the Internet a "public space" in the sense that it commands the same status as a public good? This further prompts the question whether it is an essential good to be accessed by all and at the same rate, as Wu later asserts that the Internet is "almost as necessary [as electricity] to contemporary life."

A lot of things come into play with the Internet – one must first purchase a computer or device and pay for all the costs and fees associated with the product. Designating a service as essential opens up the issue of whether the product necessary to use it is also an essential good. Companies could still provide their site to Internet users at a slower and cheaper rate, just as drivers could forgo a toll to save a buck in exchange for a longer detour. Either way, both get to the same destination.

Though consumers may benefit from faster service, eliminating net neutrality is also something that would curtail innovation and inhibit new competition from startups that do not have the funds to partner with Internet providers. A new product or service could be deferred because the firm did not have the same access to the market as a larger firm.

Whoever is responsible for and best at enforcing it, net neutrality had this going for it: it is a relatively stable, relatively open playing-field for competition. As noted, the fact that companies tried to get around it via copyright protection privileges shows that it is, in fact, doing something to enforce freedom of competition. Now, without it, there is an opportunity for concentration of power, with regard to Netflix, for example. Concentration can lead to instability, and instability leads to popular calls for state regulation, which tend in practice toward cronyism. Certainly, such a trajectory is not inevitable, but it is now more likely, giving good reason for pause at the idea that we do not need net neutrality – or something like it – in the future.

The absence of net neutrality will likely impede this bastion of ideas. Open access to the free market, including the "ultimate marketplace of ideas," is indispensable to human flourishing, and without such economic liberty the creation of wealth for all members of society may be needlessly inhibited.

Muslims can stand on both sides of net neutrality. Some Muslims may fear that faith advocacy can be curtailed by leading Internet provides who request a fee from these advocacy groups to be able to travel on the fast track. On the other hand, some Muslims may oppose net neutrality due to the fact that regulations may hinder innovations in the future.

Apart from net neutrality, various issues with regard to the technological developments have been brought to light. Regardless of whether and if so, how technology has an impact on our work, families, or faith, what we need is technology with a heart–one that beats for us.

9

Appendix A: Detailed Overview of Innovation and Competition Policies

While decisions with regard to regulations and especially policy can severely impact the economic growth, the ratio of what is known to unknown with respect to the relationship among competition, innovation, and policy is low. This section aims to provide an in-depth discussion on whether the best form of competition policy is a strong innovation policy.

9.1 Introduction

Wrong policies and the anti-competitive behavior of companies can influence the effectiveness of competition even though the markets work fair most of the time. Being one of the essential forces affecting markets, competition refers to the rivalry among companies that aim to acquire sales and earn profits (Buccirossi et al., 2013).

Although different definitions might exist for competition policy, the following definition for competition policy as defined by Buccirossi et al. (2013) will be used throughout this section: a collectivity of restrictions and requirements that constitute the essential rules of competition (or antitrust) law along with the existing set of tools of competition authorities for controlling and punishing any violation of the same rules.

9.2 Taxonomies of Innovations

There can be different determining factors for different kinds of innovations. For example, the determinants for organizational process innovations might differ from technological or product innovations. So, in order to specify these determinants, it might be useful to conduct a disaggregation based on a micro- and a meso-level of analysis of different types of innovations and establish different taxonomies for them.

As the category of innovation can be complex, it is important to understand *what* is produced by firms and *how*. As stated below, innovation can entail both process and product innovations:

- In terms of *what* is produced, product innovations can take the form of goods or services.
- In terms of *how* goods and services are produced, process innovations can be organizational or technological. Although product and process innovations may be closely interrelated, it is still crucial to make a distinction among these categories. There can be transformations among these two categories as well. To give a specific example, products that have been bought for the purpose of investment rather than immediate consumption belong to this category. While a piece of industrial equipment such as a construction machine counts as a product when it is produced it becomes a process during its use in the production process.

Table A9.1 provides a more detailed taxonomy based on these two general categories (Scott, 1980).

Table A9.1   Modes of innovation and possible policy responses based on type of market failure (Scott, 1980)

Main Mode of Innovation	Main Sources of Market Failure	Typical Sectors	Policy Instrument
Development of inputs for using industries	Financial market transactions costs facing SMEs; risk associated with standards for new technology	Software, equipment, instruments	Support for venture capital markets; bridging institutions to facilitate standards adoption
Application of inputs developed in supplying industries	Small firm size, large external benefits; limited appropriability	Agriculture, light industry	Low-tech bridging institutions (extension services)
Development of complex systems	High cost, risk (particularly for infrastructure technology)	Aerospace, electrical technology, semi-conductors	R&D cooperation. subsidies; bridging institutions to facilitate development of infrastructure technology
Applications of high-science content technology	Public nature of knowledge base	Biotechnology, chemistry, materials science, pharmaceuticals	High-tech bridging institutions

Still, some other complementary taxonomies have been used by several other scholars such as indicated by Edquist and Riddell (2000):

(1) Small ongoing incremental changes,
(2) Discontinuous extreme and major changes,
(3) Massive changes in some ubiquitous technology designed for general purpose (GPT), also referred to as "techno-economic paradigms." (Edquist and Riddell, 2000).

Although product innovations are the major driver underpinning the structure of production process innovations should also be taken into account as they are crucial for the competitiveness of firms regardless of their country or sector.

9.3 Literature Review

This section provides the theoretical background with regard to both competition and innovation policies before putting forth the theoretical results.

9.3.1 Theories Underpinning Competition Policy

From Adam Smith onward, most economists established a consensus on the workings of competition for the general interest, yet no such agreement exists with regard to the social benefit of competition policy. Some scholars, starting with the Austrian School (e.g., Von Mises, 1940), assert that any government intervention including competition policy which interferes with free markets will be for the disadvantage of the society.

Economists such as Crandall and Winston (2003) assert that the inefficiency of the antitrust law in the United States is caused by both the difficulty of making a distinction between genuine, healthy competition and anti-competitive behaviors and the underestimated market power to restrain anti-competitive abuses. They see the application of anti-trust law necessary for only blatant price-fixing and mergers to monopoly.

In disagreement with this point of view, Baker (2003) and Werden (1998) claim that competition policy has a positive net effect on social welfare. According to these scholars, social welfare is enhanced by competition policy by causing companies to give up anti-competitive conduct without any competition authority's intervention. With regard to this ongoing debate, Whinston (2006) stated that there is a gap in empirical evidence in terms of the effects of the practices prohibited by antitrust law (e.g., competitors communication on prices) and of active enforcement of antitrust law on social welfare.

The existing body of literature referred to a positive relationship among competition and productivity. By analyzing firm-level data, Aghion et al. (1997, 2001) show that productivity is spurred by product market competition. Furthermore, a theoretical framework has been provided by Aghion and Howitt (1992) to claim that productivity may be enhanced by competition-enhancing policies.

9.3.2 Theories Underpinning Innovation Policy

The analytical framework of private innovation in a market economy was specified by Schumpeter (1994), who held throughout different stages of his career two opposing ideas with regard to the interrelatedness of technological performance and market structure.

In the earlier stages of his career, Schumpeter (1994) asserted that technological advancement occurs within the research laboratories of large companies that possessed market power. Also referred to as a theory of technological contestability in modern terminology, this theory asserts that economic profits of such firms would be used to invest in large-scale and risky R&D activities that could also contribute to the society's well-being at the same time, and hence enable firms to dominate the product-market.

According to this second perspective, the level of technological progress will be greater if only a few large companies dominate the markets. Innovation investment could be undertaken better by such firms as they could take advantage of such large and perhaps diversified economies of scale, and, they would also be in a better position to apply the new technological developments in a commercially viable way. Furthermore, Schumpeter (1994) also asserted that risk is an inherent feature of commercialization and research and development; and market power provided an "insurance" against such risk.

While the debate in terms of the implications of these two positions is still ongoing, one of the recognized views was that from a social point of view, the rate of investment in R&D (research and development) is likely to be too low, regardless of the structure of the market being simple, a highly concentrated oligopoly, or in between.

In a similar vein, according to the existing body of literature of endogenous growth, technological progress is seen as a main driver of economic development. Although the accepted view that wants exceed means may hold true, the fair distribution of resources to enhancing society's means can alleviate the problems along with the globalization, economic progress, and structural adjustment. One of the most effective ways with regard to the allocation of such resources would be to target the specific kinds of market failure that prevent innovation in various industries.

Within the context of modern economy, "innovation" entails not only the formal R&D happening in research laboratories, inputs (such as funding and engineering or scientific staff), and outputs (patents) that can be analyzed and reported to government authorities, but also incremental changes in product design and development processes that happen as a result of an amalgamation of tacit, uncodified knowledge within individual firms which cannot be conveyed to or appropriated by rivals without any costs (Reinganum, 1983).

Within this light of information, good reasons exist for policy makers to encourage private innovation, and various measures can be taken by them to do so. There will be differences in such measures based on their sector and policy design which does not consider these differences across industries in terms of the sources of market failure will not be very successful.

9.3.3 The Role of Competition Policy in the Innovation Process

Understanding where competition policy does interfere with the innovation process is crucial to discuss if the best form of competition policy is a strong innovation policy.

As most of basic research does not occur in the private sector, the role of competition policy in basic research is modest. Basic research undertaken by for-private companies, especially research conducted by a single firm, is in general not limited by competition policy. Constraints of the pure research activities due to competition policy might arise when two or more firms conduct joint research (D'aspremont et al, 2000). Competition authorities in general stated that they will not interfere with research done by research joint ventures (RJVs) and other forms of collaboration. To give a specific example, in the United States, legislation was passed expressly to ease antitrust limits on RJVs and there are virtually no examples of joint research activities that have been blocked because of antitrust regulations. In general, the concept that basic research entails aspects of a public good, and hence is very appropriate for RJVs, is a commonly accepted idea in antitrust circles (D'aspremont et al, 2000). Specifically, collaboration is accepted as the most relevant means to undertake so-called "pre-competitive" research, i.e., research that can effectively be used and shared by several industry participants, so that firms can later compete by developing their own unique products (D'aspremont et al, 2000).

Competition policy does also not play a very active role in the invention process. Although in the United States, there is a mention of "innovation markets" in the Department of Justice (DOJ) Guidelines which is prepared by Federal Trade Commission (FTC) with regard to licensing of Intellectual Property ("IP Guidelines"), this type of analysis relates to traditional issues of potential competition.

Whether or not competition policy has an important influence on the *incentives* for private firms in order for them to incur the costs necessary to produce inventions is another question, yet these impacts occur due to the rules putting a limit onto the subsequent behavior and returns throughout the commercialization, diffusion, and extensions phases of the innovation process (D'aspremont et al, 2000).

9.3.4 Competition in the Market or for the Market?

Within the context of innovative industries, competition can be thought of as a number of races for the development of new technologies (D'aspremont et al, 2000). Once a leadership position is attained, victory is achieved in a single product market or more. Yet, this should not be interpreted as if the winner can then have a rest to enjoy the victory. In order to maintain leadership, another new race should be entered immediately (D'aspremont et al, 2000).

Due to the winner-takes-most or all outcomes and the technological opportunity, the competition form that matters most from a welfare perspective is the competition *for* the product market rather than the competition *in* a product market such as the case of advanced industries. An example for the prior one is a race for being the first deliverer of a new product to market or the first producer by use of a new technology. Contrary to advanced industries in which new entrants receive the market share on a gradual basis, the dominant incumbent is rapidly replaced as a result of successful entry in such innovative industries (D'aspremont et al, 2000).

Due to the substantial R&D cost in these sectors that are mostly not dependent on volumes of production, the total cost becomes less on average with output. Therefore, companies cannot continue to exist by offering an amount close to the marginal cost of production (D'aspremont et al, 2000).

Moreover, effects of both network and positive feedback should be taken into account in knowledge-based sectors. Network effects occur due to the increase in value attributed to a particular product by each user as the total user number increases (D'aspremont et al, 2000). Positive feedback exists when goods complement each other and widened use of that particular product enhances the value of other products more (D'aspremont et al, 2000). To give a specific example, a larger user base for an installed operating system increases the value of a spreadsheet software running on that system. Similarly, due to the widened use of that software, the operating system's value also increases. Both of these positive feedback and network effects lead to increasing returns to scale on an inter-temporal basis (D'aspremont et al, 2000). Markets in which these effects are strong experience "tipping" (D'aspremont et al, 2000), in other words, as time passes a single product can dominate due to the fact that there might be an unstable existence of various incompatible products. As a result of tipping, a *de facto* standard will be adopted (D'aspremont et al, 2000).

Within the innovative industries, competition policy can affect welfare through the following channels (D'aspremont et al, 2000):

- The race's intensity to make an innovation.
- Competition within the context of the product market, with a high probability to influence variables such as price and service quality.
- Competition within the context of the licensing market.

Behavior which reduces competition in one field may increase it in another. As a consequence, competition policy is exposed to two interrelated yet conceptually different issues (D'aspremont et al, 2000):

i) How to evaluate the impacts on competition in each of the previously mentioned three fields;
ii) How to calibrate for the trade-off between opposite impacts of competition in different fields; specially, how to weigh both static and dynamic factors of competition.

Within the context of innovative industries, these characteristics of competition relate to the question of how one evaluates the success of competition policy. If the consideration is that competition policy be inclined toward the aspects of competition *in* the product market, then its success depends on whether dynamic efficiency is increased by the product market's competition at a more intense level (D'aspremont et al, 2000). So, if there are crucial races to reduce costs or to deliver new products to market, and if winners are expected to acquire most or all of the market share, then competition policy should be evaluated on the ground of whether it enhances the incentives to make an innovation (D'aspremont et al, 2000).

9.4 Theoretical Results

Throughout the literature, several factors such as market failure, limited appropriability, and external benefits to knowledge production imply that pure reliance on a market system will lead to an underinvestment in the field of innovation, in comparison to its socially desirable level, so a strong innovation policy may not always be the best form of competition policy. First, theoretical results are put forth and then both empirical and case-study evidence is provided before making recommendations on policy formulation.

9.4.1 Competition versus Monopoly

In order to analyze the issues raised by Schumpeter (1994), Arrow (1962) developed a simple framework to make a comparison between the monopolist and a single entity (operating in similar markets except for their supply structures) in terms of their profitability of a cost-saving innovation in R&D and production.

The uncertainty or imperfect appropriability of the profit produced by the successful innovation was not taken into account by Arrow (1962). Apart from crucial aspects of the R&D process, time and strategic interactions in R&D among rival suppliers were also left aside as only the extreme cases of a single entity (such as monopoly) and several small companies were considered. As a simple comparison between the monopolist and the individual rival company in terms of the benefit of a cost-saving innovation, the cost aspect of the R&D process or the financing of R&D was also not considered while keeping the demand side of the market the same in both cases.

Within the context of his model, Arrow (1962) asserted that the profit of the competitive firm from successful innovation would exceed the monopolist's profit. From the point of view of the monopolist, the incentive to make an innovation occurs due to the incremental financial gain from a production at lower unit cost instead of higher unit cost. Part of the financial gain made after innovation by the monopolist just replaces the financial gain made before innovation. According to Arrow's model, an entity in a perfect competitive sector will gain more from a successful innovation in comparison to an entity in an otherwise similar monopolized sector (Arrow, 1962).

9.4.2 Racing Models

Some models within the literature such as those developed by Lee and Wilde (1980) consider the development time as a random variable. A company can decrease the *expected* discovery time by making more effort for R&D or increasing its R&D budget, but the actual time for discovery is dependent on luck.

The early stream of work in the racing literature considered the income occurring due to the successful innovation to be exogenous, and investigated how the expected discovery time might alter as the total amount of independent R&D projects gets bigger. Results depend on details of how the cost function of R&D was specified, yet in general it was argued that the expected discovery time decreases as the amount of entities conducting R&D projects increases.

Considering of R&D as a random process weakens the argument that it is inefficient for various companies to follow the same research goal (Nelson, 1961). As several R&D projects with the aim of developing a new product or process might pursue different strategies, it is not possible to know in advance which of these strategies will prove to be more successful. Due to the existence of several R&D projects with the same goal the probability of success of at least one project is increased. As the successful project cannot be known in advance, it will not be useful to make a retrospection after innovation and label the unsuccessful projects as being wasteful.

The later part of the literature with regard to racing (Gilbert and Newbery, 1982; Reinganum, 1983, 1984) investigates if a potential entrant or an incumbent company could make an innovation earlier, in the sense of an expected value. These frameworks provide a generalization of Arrow's model and they don't make any separation between R&D cooperation and the product-market cooperation.

The path-dependence literature suggests that there can be more preference for the next or later versions of an innovation, yet these may be precluded from successful implementation due to time constraints. This argument is against the very early emphasis on a specific line of development.

Although details of model development shape the specific results, the results suggest that in comparison to large incumbents, smaller companies and potential entrants are likely to invest more in innovation. This view is also in line with the innovation perspective of "creative destruction" rather than the perspective of product market innovation.

9.4.3 R&D Cooperation Policy

Local antitrust or competition policy, generally, favors competition as entities make their own independent decisions and follow their self-interest in contrast to the international "competition" policies such as promotion of national champions.

According to this presumption, two assumptions underpinning R&D process are:

- Private investment in innovation on its own will not reach the socially desirable level.
- Adopting a permissive competition policy attitude toward R&D cooperation will enhance technological performance.

Viewed from a theoretical perspective, although R&D cooperation may enhance the performance of the market, this may not necessarily happen due to increased levels of innovative effort or output. On the contrary, although innovative effort may get reduced in such models, the market performance in general increases. Due to a joint R&D, various firms will be able to access the new products or technology in the post-innovation market. From a consumer perspective, this is good as greater product-market competition is ensured after discovery rather than a single company controlling the innovation. It is exactly this enhanced competition that decreases the incentive for a cooperative project to make an R&D invest investment. So, although there is a delay in innovation (in terms of an expected value), users are much better off after the discovery happens.

9.4.4 R&D Subsidies

Contrary to the R&D cooperation that deals with knowledge spillovers in a direct way, by enabling the coordination of R&D activities for companies that gain from spillovers, subsidies' way of dealing with spillovers is indirect: while spillovers decrease the marginal benefit of a successful innovation, subsidies can cause the marginal cost of innovation investment to be decreased, if well designed. As the basis for companies to make their decisions is their own profitability, the additional gain to users of additional R&D is not taken into consideration. The government can influence through the means of subsidies the companies' decisions in such a way that this external benefit to consumers requires. Yet, subsidies on their own might not be enough to eliminate underinvestment in R&D and the market failure (Scott, 1980).

The government can also influence through the means of subsidies the structure of R&D cooperation. Due to the existence of a subsidy, a cooperative R&D can also occur whereas companies, if left to their own devices, would prefer to conduct R&D independently. Subsidy may lead to a joint venture, which is one of the crucial forces for growth and innovation in the early years within a particular industry.

9.4.5 Patents

Despite the general understanding that patents may lessen the incentives for companies' R&D investment, patents increase the probability for a product-market competition among patent-holders over other products or processes which can be substituted (Steinmueller, 1987). Obtaining a patent increases the incentives for the patent-holder to provide a license for the patent use in order to avoid cases of "inventing around" the patent. In this way, the diffusion of the patented innovation can also be facilitated (Steinmueller, 1987).

The patent's breadth and length are interrelated. Although patents decrease the incentive for R&D investment, this can be mitigated by increasing the patent coverage's length. Matutes et al. (1996) assert that a narrow patent regime may cause delays in patenting, as innovators might privately develop business applications before the disclosure of the innovation's traits to their rivals. The rate of technological progress is reduced due to this delay of disclosure. Although the scope of patent protection should be broad enough to foster initial R&D investment, it should not be so broad as to hinder the search for other improvements.

Levin et al.'s (1984, 1987) claim that as appropriability mechanisms, in most R&D intensive sectors, lead time, secrecy, learning-curve effects, and efforts for sales and service received a higher ranking than patents. According to Vonortas (1994), the underlying reason is the significant "learning-by-doing" investment necessary to use information. Technological knowledge entails a combination of both incomplete and poorly defined know-how and a highly codified information set which is hard to obtain (Vonortas, 1994). So there will be some degree of appropriability of the income occurring because of the innovation, regardless of the nature of the patent system (Vonortas, 1994).

Depending on the sector, the role of patents may also differ. Mowery (1995) states that throughout the history of the US semiconductor industry, patents acted as bargaining chips that allow a firm the trade of licenses to access technology protected by other patent-holders. In these cases, patents lead to the diffusion of technology. One example is the Semiconductor Chip Protection Act of 1984, which facilitated patent protection for computer chip designs and hence increased the possibility for patent infringement suits (Mowery and Rosenberg, 1993).

Although technological progress can be fostered by the patent system by increasing post-innovation appropriability, this will not always happen given the presence of extensive knowledge externalities. To give a specific example, based on a 1956 antitrust consent decree, AT&T required licenses for a patent-controlled technology at suitable rates and it required cross-licenses of patents from companies to which licenses were granted. This pattern of cross-licenses enabled the fast entry into and development of the US semiconductor chip industry (Steinmueller, 1987). The knowledge diffusion in this industry was also facilitated by staff's free movement among rival companies, including startups financed in venture capital markets.

9.5 Empirical Evidence

There is substantial evidence based on case studies that despite the fact that technological frontiers may be pushed out by joint R&D, the main advantage of such an R&D occurs due to the information diffusion among stakeholders and product-market competition following the diffusion. Due to the product market rivalry, the incumbents try to adopt state-of-the-art developments on time, and as a result technological performance is promoted.

In case of an extremely high R&D cost, subsidies for single or cooperative R&D may be effective. When granting direct public subsidies, an agreement should be made by recipients that the research results will be openly licensed at reasonable rates.

According to the rent-seeking literature (Fudenberg et al., 1983; Fudenberg and Tirole, 1987; Anderson et al., 1997), overinvestment in innovation is also possible as rivals that strive to gain a post-innovation competitive advantage may have from the perspective of the society a higher spending in total. This is referred to as "the overbidding problem" by Baldwin and Scott (1987) in contrast to the "appropriability problem" that causes underinvestment.

9.5.1 The Rate of Return to Investment in R&D

As the existing body of literature suggests, the social rate of return to investment in R&D can be estimated by means of measurement of some R&D input or output as a variable of an explanatory equation that relates to the increase in total factor productivity (Jones and Williams, 1997). The approximate R&D coefficient shows how much productivity growth would be increased if R&D activity were increased.

According to Jones and Williams (1997), the social rate of return to R&D is made of various components that demonstrate the value of additional ideas *per se*, the value of the additional output enhanced by increased knowledge, and the value of increased spillovers.

Still some other studies include evidence about factors that have an effect on the rate of return to investment in R&D. In some studies such as Link's research (1987) publicly funded R&D is taken into account as a separate constituent of productivity; so the impact of public funds with regard to the R&D's rate of return can be estimated. According to Link (1987,1997),

(1) In his sample of manufacturing companies, the productivity growth for government support of basic research exceeded the one for government support of applied R&D;

(2) In general, the availability of government funding reduces the approximated private rate of return on R&D.

One commonly accepted view is that public funding provides support for private companies to do projects that are socially advantageous but less privately advantageous than the average projects in portfolios of the private firms' R&D (Link (1987, 1997).

9.5.2 Cross-sectional Evidence

Another line of research investigated the question of whether or not government funding removes private spending which has been asked by Irwin and Klenow (1996) by making use of cross-sectional evidence. One conclusion reached by Scott (1980) and Mansfield (1985) was that public funding does not dislocate private R&D spending; instead the private spending grows to some extent from what it would have happened without the public funds. In a large cross section including more than 400 firms with over 3,000 observations on their activities in several industries as done by Scott (1980), it was observed that for each federal research dollar earned by the average company, the company spent an additional eight one-hundredths of a dollar of its own money on R&D, other things being held equal, including a total set of industrial effects.

In order to interpret these different findings with regard to the dislocation of private spending by public spending, one question that might be asked is whether it can be specified which cooperative and/or public R&D acts as either a substitute for or a complement to private and independent R&D.

In alignment with the empirical literature on public funding of R&D investment, one answer is that public R&D generally complements private R&D undertaken by small firms (Acs et al., 1994a,b; Lerner, 1995; 1996).

Moreover, Lichtenberg's (1987) focus on the relationship between public and private R&D and private sales versus sales to the government could offer an explanation with regard to the complementarities found for the larger firms, rather than concluding that the gain of smaller firms from the government funding to contribute to individually performed R&D investments would probably be higher.

In light of this information, private spending is not dislocated by public funding of R&D *in general* except in a few specific cases.

9.6 Suggestions

To make use of the insights suggested by modern industrial economics, which is a collection of unique game-theoretic models for specific markets that entail detailed assumptions about order of moves, strategic variables, data sets, and solution concepts, policy advice on innovation should be tailored. This section takes as its starting point the general agreement within the existing body of literature that from a social perspective, relying purely on market processes may result in an underinvestment in R&D related to innovation. So R&D and innovation policy should take as its basis specific sources of market failure in order to be designed effectively for competition rather than focusing on problems that do not affect a particular industry.

If an open economy and innovation includes the competition policy as one of its parts, both domestic investment and FDI can be fostered by means of this policy, as it contributes to the investor confidence by offering a consistent framework within the related sector. Not only does an effective competition policy encourage innovative new entrants to take a crucial role in the development process, but it also contributes to development while reducing opportunities for corruption and rent seeking.

When designing an appropriate national competition policy both the local realities including the governance capabilities and political realities including the presence of small and frequently vulnerable domestic markets should be taken into account in order to enable a strong innovation policy. A culture of competition should be created among government, firms, and consumers by gaining an increased understanding of the objectives of innovation policy. Civil society and an open media specifically can act as a constructive and valuable means for building a culture of competition. Above all, politicians must be committed to having a detailed understanding of their official duties toward markets and to supporting the building of the technical capacity required for this task.

9.7 Conclusion

Fair competition plays a crucial role both for reducing poverty and economic development. Eliminating unnecessary distortions to competition can facilitate significant reforms within innovation policy. A detailed understanding is essential at innovation policy levels by not only government or the business sector, but also consumers, in terms of the positive impact of both effective competition and competition policy on an economy.

10

Appendix B: Conceptualizing Cyber Security from the EU Perspective

In 2004, ENISA (European Network and Information Security Agency) was founded in order to facilitate "best" practice among Member States with regard to cyber security policies with regard to EU's Information Society agenda. In 2007, due to DDoS (Distributed Denial of Service) attacks on the public infrastructure of Estonia, the EU along with NATO and other related actors considered changing their approach. As a result, the EU's policy has been developed within the light of the Europe 2020 strategy. The European Commission (2013) and its Principles and Guidelines for Internet Resilience and Stability (2011) focus on the significance of global partnerships to address both military and civilian aspects of cyber security challenges.

According to the EU Cyber security strategy, the Internet must be kept protected and "open and free" based on the same values and principles that the EU considers for offline space (EU Cyber security Strategy). EU Cyber security Strategy and its Directive on Network and Information Security (NIS Directive) were published on 7 February 2013 to require the reporting of significant cyber incidents across all critical infrastructure sectors (NIS Directive, 2013). For the first time, the EU tried to specify priorities with regard to the protection of cyberspace via means of this strategy as previously there was no coordination with regard to the construction of an effective security ecosystem for cyberspace (Klimburg and Tirmaa-Klaar 2011).

This section will be structured as follows: After a brief overview of the conceptual landscape of the EU's cyber security development, the suggested tools for cyber security policy of the EU will be explained. Next an overview of the EU's way of dealing with cyber security threats will be explained. The final section will provide recommendations for the EU's cooperation with other states on cyber security in the future.

10.1 Concepts and Approaches of the EU's Cyber Security Policy

The existing body of academic literature with regard to the EU's action in cyber security is scarce as most of the available work focuses on the United States and other regions (Kshetri, 2013), with no in-depth theoretical analysis of the EU's cyber security policy. Various approaches such as managerial and strategic (Libicki 2007, 2009; Clarke and Knake 2010), historical (Carr, 2009) and other approaches that focus on terrorists (Wiemann 2006; Colarik 2006) have been used. While the emphasis of such approaches has been more on recent cyber threats and how to establish the "cyber peace" (Clarke and Knake 2010), other theoretically and methodologically driven works used innovative mixed-method (Deibert et al. 2011), regulatory (Brown and Marsden, 2007) and other approaches that try to evaluate the extent of securitization of cyber policy (Dunn, 2007, 2008; Bendrath et al. 2007).

Cyber power has been so far one of the most frequently used concepts with regard to cyber security (Klimburg and Tirmaa-Klaar 2011; Betz and Stevens 2011; Klimburg 2011; Nye, 2010; Kramer et al. 2009). While Nye (2010) defines cyber power as the ability to utilize the digital pace to create an influence and gain advantages in other operational contexts (2010, p. 4) he makes a distinction between information and physical instruments, as well as soft and hard power in cyber space, and provides examples of how they can be used both outside (extra-cyberspace power) and inside (intra-cyberspace power) (See Table A10.1).

Other scholars such as Betz and Stevens (2011, p. 44) acknowledge the fluidity of cyberspace and mention that and that various non-state and state actors, ranging from states to citizens, global networks, and organizations, can have an influence at any point in time in order to exploit the possibilities

Table A10.1   Instruments of power in cyberspace

	Intra-cyber Space	Extra-cyber Space
Information instruments	Soft: Set standards and norms	Soft: Public campaign to influence opinion
	Hard: Denial of Service Attacks	Hard: Attack SCADA systems
Physical instruments	Soft: Infrastructure to support activists of human rights	Soft: Protests to name and shame cyber providers
	Hard: Government controls over enterprises	Hard: Cut cables or bomb routers

Source: Nye, 2010.

offered by cyberspace. As a result, they conceptualize the cyber power in four distinct forms:

- *Compulsory:* This occurs when one cyberspace actor makes use of direct coercion to change the behavior of another actor (hard power such as attacks on FBI systems);
- *Institutional:* This refers to one of the actor's Indirect control by means of formal and informal institutions (soft power such as setting norms);
- *Structural:* This refers to a control type which aims to maintain the existing structures and enable or limit the actions of the actors with regard to those with whom they have a connection (soft power regarding how cyberspace can enable or limit the actions of actors);
- *Productive:* This refers to the definition of the "fields of possibility" that limit or enable social action (soft power such as a states' construction of the "hacker" as threat) by means of discourse.

According to Klimburg (2011), cyber power has the following other crucial aspects:

1) Coordination of policy and operational aspects across governmental structures;
2) Policy coherency through international alliances and legal frameworks;
3) Cooperation among non-state cyber actors.

In opposition to Nye (2010), Klimburg (2011) asserts that the last one is the most significant one as most of the control is exerted by civil and business society due the nature of the cyberspace and the state's capability is restricted to indirect influence. Based on the Model of Integrated Capability (Klimburg and Tirmaa-Klaar 2011), Klimburg (2011) emphasizes the need for a holistic approach to cyber security, in other words a soft-power approach which is focused on the inward is fundamental for the creation of a "whole of nation" of cyber capability' (2011, p. 43). Although the European Commission (2013) included some of these scholars' recommendations, it is not certain whether "hard" cyber power considered in EU's conventional national security terms is actually aligned with the EU's core values. This is a crucial point given the revelations of post-PRISM (e-spying) that certain EU Member States such as Sweden and the United Kingdom were complicit in mass data surveillance of citizens in contravention of the EU laws on data protection among "friends" in Europe.

Within the light of this information, it can be argued that the EU needs a security policy based on resilience approach (Christou, 2015) and a specific kind of soft cyber power based on Klimburg's three aspects including productive and institutional cyber power, rather than the conventional and hard cyber power often exercised by democratic and authoritarian states, based on the logic of cyber sovereignty. In the post-Snowden era, this is even more imperative as the consequences of a mindset of "national security first" have been in sharp contrast with the vision of EU's open, safe, and secure cyberspace. In this regard, Dunn-Cavelty (2013, p. 3) states that a special type of "soft" power driven by core values and internal resilience should be developed by the EU to make sure that its normative vision with regard to the cyberspace's governance is obtained within the global arena.

10.2 The EU Approach to Cyber Security

Over time as the EU's approach to cyber security evolved in an *ad hoc* and fragmented manner, several strands of cyber security "policy" emerged which can be categorized as in Table A10.2:

According to Robinson (2013), given the complexity of the cyber security domain, and the difficulty for the design of a coherent internal EU policy, these strands entail the following aspects:

- Legal (enforcement),
- Economic (Internal Market)
- Security (CSDP).

Furthermore, the Cyber security Strategy also includes both the development of technological resources for cyber security and the establishment of policy for cyberspace in order for the EU to contribute to its the fundamental values. (Robinson, 2013):

Table A10.2   Strands of EU's cyber security policy

Policy Strand	Responsible Institution
Cybercrime and cyber attacks	Directorate General Justice and Home Affairs
Network and Information Security (NIS): a) Critical Infrastructure Protection (CIP) b) Critical and Information Infrastructure Protection (CIIP)	Directorate General Connect (previously DG Information Society)
A cyber defense element	European External Action Service (EEAS)

Source: Robinson, 2013.

Based on the normative foundation of the EU approach, the global Internet is regarded as a collective or public good which should be made accessible and available to all (European Commission, 2013; European Principles and Guidelines, 2011). So the use of the Internet should only be constrained when instruments and measures are used to provide harm to others. An effective cyber security strategy should take as its basis the main norms as stated in EU's Charter with regard to its Fundamental Rights (European Commission, 2013). Thus, the main EU values and norms are at the core of both online and offline activity, as stated in its Cyber security Strategy.

According to the European Commission (2013), the governance model of its cyber security policy should be based on multi-stakeholderism which arises due to the complex interactions among several actors using and managing the Internet. Some states such as Iran, Russia, China, or India held the opinion that due to the excessive power of the United States over the management of the Internet these countries themselves are under-represented in current Internet governance institutions such as the ICANN (Internet Corporation for Assigned Numbers and Names) or IGF (Forum for Internet Governance).

As any of these principles on EU cyber security policy cannot exist in a vacuum, a specific type of public–private partnership is supported under this multi-stakeholder umbrella. Therefore, while the appropriate forms and modes of governance (i.e., incentives) should be agreed by public authorities in consultation with related stakeholders, the private sector plays an important role in day-to-day governance of the Internet (European Principles and Guidelines, 2011). If there is any global disagreement with regard to the role of technical standards, data protection, who should regulate the Internet, and the appropriate legal conventions for fighting cybercrime (e.g., the Budapest Convention), these can undermine any effort to develop a secure cyberspace for everyone. In the post-Snowden era, an increased support for the Governmental Advisory Committee in ICANN was provided by the EU, by assigning it a greater decision-making role in policy on Internet governance.

10.3 Progress and Challenges of the EU Cyber Security Strategy

As Robinson (2014) states, in comparison to the United States, China, or Russia, whose main focus is on a national security (threat) logic and hence deterrence and militarization constitute their central strategy (hard power), the EU is focused on building resilience to enable fast recovery from cyber attacks, developing the required capacities to resist cyber attacks, and fighting cybercrime (soft power). For the EU, those priorities that focus on non-military aspects which aim to create partnerships for the development of an effective cyber security culture within and beyond the EU are more important that its military and intelligence infrastructures (Bendiek, 2014). include not only dialog, incentives, platforms for cooperation and coordination, and voluntary arrangements (to ratify the Budapest Convention), but also more formal requirements, such as the suggested NIS Directive to protect both the privacy and data of its citizens. Yet, it is not known which of these instruments provide the EU with the opportunity to create effective productive (soft) and institutional cyber power.

With regard to the preparation of its Cyber Defense Policy Framework, the EU so far developed bilateral relationships by means of participation in international platforms such as the London conference and appropriate international fora (OECD, ICANN, ITU, IGF), as well as international organizations such as NATO.

On the other hand, there have been some issues with the CoE (Budapest) Convention on Cybercrime as it was not implemented by all Member States across Europe. Besides, some countries such as Russia and China oppose ratifying the Convention as they are concerned that national security culture would be undermined because Internet users in these countries are often the alleged source of many cyber security attacks (Goldsmith, 2011).

Being members of SCO (Shanghai Cooperation Organization), Russia, China, Tajikistan, and Uzbekistan proposed in September, 2011 that the UN Secretary General should establish a dialog about the new "Code of Conduct for International Security" with the purpose of reaching an agreement on international norms of behavior for the Internet. Although at first sight, its main principles (Appendix C) are in alignment with what is being advocated by the EU with regard to full respect for rights and freedoms in cyberspace, an in-depth look reveals that rather than supporting a multi-stakeholder approach to cyber security, the commitment to prevent other states from using core technologies to threaten other countries' security is underpinned by a sovereign logic of control rather than freedom (Gjelten, 2010).

The promotion of the UN to take a more active role in Internet governance, where China has much more influence and weight in terms of voting behavior, could lead to a potential separation of national Internet spaces, rather than multi-stakeholder governance. In addition to the perceived dominance of the United States in controlling the Internet, given the SCO members' commitments to restrict information flows in case of any impact upon their security culture, the main concerns are about repressive regimes' excuses to restrict access to independent external news sources, as in the case of the Arab uprisings (Gjelten, 2010). As some scholars argue, this might also be used as a justification to develop draconian laws on access and dissemination of information which might threaten "national security" (Healey 2011b; Bendiek, 2014). Finally, the fact that states such as Russia and China which are seen to be at the center of cyber conflict do not support a commitment to control patriotic hackers is regarded as another concern. Besides, no Code calls for all states to sign up and be bound by the laws of armed conflict which raises many questions with regard to the legitimacy and proportionality of certain targets (Christou, 2015).

10.4 Conclusion

If the aim of the EU is the establishment of deeper cooperation with other nations within the context of cyber security in the future, platforms (e.g., the Task Force) should create an effective agenda that reflects the differences between the EU (soft power) and other countries such as China or Russia (hard power). Yet, there should not be any compromise in the principles and norms of these countries with regard to their Internet policies. Although this may sound as being too difficult to accomplish it is not impossible given EU's increased emphasis on cyber security along with its evolving cyber security strategy.

11

Appendix C: International Code of Conduct for Information Security (SCO)[1]

1. To comply with the UN Charter and universally recognized norms governing international relations, which enshrine, inter alia, respect for the sovereignty, territorial integrity and political independence of all states, respect for human rights and fundamental freedoms, as well as respect for diversity of history, culture, and social systems of all countries.
2. Not to use ICTs including networks to carry out hostile activities or acts of aggression and pose threats to international peace and security. Not to proliferate information weapons and related technologies.
3. To cooperate in combating criminal and terrorist activities which use ICTs including networks, and curbing dissemination of information which incites terrorism, secessionism, and extremism or undermines other countries' political, economic, and social stability, as well as their spiritual and cultural environment.
4. To endeavor to ensure the supply chain security of ICT products and services, prevent other states from using their resources, critical infrastructures, core technologies, and other advantages, to undermine the right of the countries, which accepted this Code of Conduct, to independent control of ICTs, or to threaten other countries' political, economic and social security.
5. To reaffirm all states' rights and responsibilities to protect, in accordance with relevant laws and regulations, their information space and critical information infrastructure from threats, disturbance, attack and sabotage 6. To fully respect the rights and freedom in information space, including rights and freedom of searching for, acquiring and disseminating information on the premise of complying with relevant national laws and regulations.

---

[1] *Source:* http://news.dot-nxt.com/2011/09/13/china-russia-security-code-of-conduct

6. To promote the establishment of a multilateral, transparent, and democratic international management of the Internet to ensure an equitable distribution of resources, facilitate access for all, and ensure a stable and secure functioning of the Internet.

7. To lead all elements of society, including its information and communication private sectors, to understand their roles and responsibilities with regard to information security, in order to facilitate the creation of a culture of information security and the protection of critical information infrastructures.

8. To assist developing countries in their efforts to enhance capacity-building on information security and to close the digital divide.

9. To bolster bilateral, regional, and international cooperation, promote the United Nations' important role in formulation of international norms, peaceful settlement of international disputes, and improvement of international cooperation in the field of information security, and enhance coordination among relevant international organizations.

10. To settle any dispute resulting from the application of this Code through peaceful means and refrain from the threat or use of force.

References

[1] Acs, Z. J., Audretsch, B., and Feldman, M. P. (1994a). R&D Spillovers and innovative activity. *Manage. Decis. Econ.* 15, 131–138.

[2] Acs, Z. J., Audretsch, B., and Feldman, M. P. (1994b). R&D Spillovers and recipient firm size. *Rev. Econ. Stat.* 126, 336–340.

[3] Aghion, P., Harris, C., Howitt, P., and Vickers, J. (2001). Competition, imitation and growth with step-by-step innovation. *Rev. Econ. Stud.* 68, 467–492.

[4] Aghion, P., Harris, C., Howitt, P., and Vickers, J. (1997). Competition and growth with step-by-step innovation: an example. *Eur. Econ. Rev.* 41, 771–782.

[5] Aghion, P., and Howitt, P. (1992). A model of growth through creative destruction. *Econometrica* 60, 323–351.

[6] Anderson, S. P., Goeree, J. K., and Holt, C. A. (1997). *Rent Seeking with Bounded Rationality: an Analysis of the All-Pay Auction. Thomas Jefferson Center Discussion Paper* 283, Charlottesville, VA: University of Virginia.

[7] Arrow, K. (1962). "Economic welfare and the allocation of resources for Invention," in *The Rate and Direction of Incentive Activity Economic and Social Factors*, ed. R. Nelson (New Jersy, NJ: Princeton University Press).

[8] Baker, J. B. (2003). *Why did the Antitrust Agencies Embrace Unilateral Effects*? Virginia, VA: George Mason University Law Review, 12, 31–38.

[9] Baldwin, W. L., and Scott, J. T. (1987). *Market Structure and Technological Change*, in the Series *Fundamentals of Pure and Applied Economics,* Vol. 17, New York, NY: Harwood Academic Publishers.

[10] Bendiek, A. (2104). 'Tests of Partnership: Transatlantic Cooperation in Cyber Security, Internet Governance and Data Protection', SWP Research Paper, RP5, March 2014, Berlin.

[11] Bendrath, R., Eriksson, J., and Giacomello, G. (2010). "From 'cyberterrorism' to 'cyberwar', back and forth: how the United States securitized

cyberspace," in *International Relations and Security in the Digital Age* eds J. Eriksson. G. Giacomello. (New, NY: Routledge).

[12] Betz, D., and Stevens, T. (2011). *Cyberspace and the State: Towards a Strategy for Cyber-Power, the International Institute for Strategic Studies*. Oxon, Routledge.

[13] Brown, I., and Marsden, C. T. (2007). *Co-regulating Internet Security: the London Action Plan*. Available at: http://essex.academia.edu/ ChrisMarsden/Papers/700007/Co-regulating_Internet_security_the_ London_Action_Plan [accessed April, 2016]

[14] Buccirossi, P., Ciari, L., Duso, T., Spagnolo, G., and Vitale, C. (2013). Competition policy and productivity growth: an empirical assessment. *Rev. Econ. Stat*. 95, 1324–1336.

[15] Carr, J. (2009). Inside Cyber Warfare: Mapping the Cyber Underworld. O'Reilly Media.

[16] Christou, G. (2015). "Cyber Security in the European Union: resilience and adaptability," in *Governance Policy, New Security Challenges Series*, Vol. 92, Basingstoke: Palgrave Macmillan.

[17] Clarke, R. A., and Knake, R. K. (2010). Cyber War: The Threat to National Security and what to do about it. New York, NY: Harper Collins.

[18] Crandall, R. W., and Winston, C. (2003). Does antitrust policy improve consumer welfare? assessing the evidence. *J. Econ. Perspect*. 17, 3–26.

[19] D'aspremont, C., Encaoua, D., and Ponssard, J. P. (2000). "Competition policy and game theory: Reflection based on the cement industry case", in *Competition Policy and Market Structure: Game Theoretic Approaches, eds* J. Thisse and V. Norman (Cambridge, Cambridge University Press).

[20] Deibert, R. J., Palfrey, J. G., Rohozinski, R., and Zittrain, J. (2011). *Access Contested: Security, Identity and Resistance in Asian Cyberspace*, Vol. 12, Cambridge: MIT Press.

[21] Dunn, C. M. (2007). Cyber-terror-looming threat or phantom menace? the framing of the us cyber-threat debate. *J. Info. Technol. Pol*. 4, 19–35.

[22] Dunn-Cavelty, M. (2008). *Cyber-Security and Threat Politics: US efforts to secure the Information Age*. New York, NY: Routledge.

[23] Dunn-Cavelty, M. (2013). 'A Resilient Europe for an Open, Safe and Secure Cyberspace', Occasional Papers, No.23, The Swedish Institute of International Affairs.

[24] Edquist, C., and Riddell, C. (2000). "The Role of Knowledge and Innovation for Economic Growth and Employment" in *the IT Era*, eds

K. Rubenson and H. Schuetze (Transition to the Knowledge Society, Ottawa).

[25] European Commission (2013). 'Cybersecurity Strategy of the EU: An Open, Safe and Secure Cyberspace', Brussels, 7.2.13, JOIN (2013) 1 FINAL.

[26] European Commission (2009). 'Internet Governance: Next Steps', Communication from the Commission from the European Parliament and the Council, Brussels, 18 June, COM (2009) 277 final.

[27] European Cybercrime Centre – one year on (2014), European Commission, Press Release, Brussels, IP/14/129, 10 February 2014.

[28] European Cybercrime Centre: *First Year Report*, Available at: https://www.europol.europa.eu/content/european-cybercrime-center-ec 3-first-year-report [accessed 24 April 2016]

[29] European Principles and Guidelines for Internet Resilience and Stability, March 2011. Available at: http://ec.europa.eu/danmark/documents/ alle_emner/videnskabelig/110401_rapport_cyberangreb_ en.pdf [accessed 11 April 2016]

[30] Fudenberg, D., Gilbert, R., Stiglitz, J., and Tirole, J. (1983). Preemption, leapfrogging and competition in patent races. *Eur. Econ. Rev.* 22, 3–31.

[31] Fudenberg, D., and Tirole, J. (1987). Understanding rent dissipation: on the use of game theory in industrial organization. *Am. Econ. Rev.* 77, 176–183.

[32] Furnell, S., and Warren, M. (1999). 'Computer Hacking and Cyber Terrorism: The Real Threats in the New Millennium.' *Computers and Security* 18(1).

[33] Garrido, M., and Halavais, A. (2003). "Mapping networks of support for the zapatista movement: applying social networks analysis to study contemporary social movements," in *Cyberactivism: Online Activism in Theory and Practice* eds M. McCaughey and M. D. Ayers (London: Routledge).

[34] Gilbert, R. J., and Newbery, D. M. G. (1982). Preemptive patenting and the persistence of monopoly. *Am. Econ. Rev.* 72, 514–526.

[35] Healey, J. (2011a). *Comparing Norms for National Conduct in Cyberspace*. Available at: http://www.acus.org [accessed March 12, 2016].

[36] Healey, J. (2011b). *Breakthrough or Just Broken? China and Russia's UNGA Proposal on Cyber Norms*. Available at: http://www.acus.org [accessed April 12, 2016].

[37] Irwin, D. A., and Klenow, P. J. (1996). High-tech R&D subsidies estimating the effects of sematech. *J. Int. Econ.* 40, 323–344.

[38] Jones, C. I., and Williams, J. C. (1997). *Measuring the social return to R&D*. Finance and Economics Discussion Series Staff Working Paper 12. Washington, DC: Federal Reserve Board.

[39] Lee, T., and Wilde, L. L. (1980). Market structure and innovation: a reformulation. *Q. J. Econ.* 94, 429–436.

[40] Lerner, J. (1995). Venture Capitalists and the Oversight of Private Firms. *J. Finance* 50, 301–318.

[41] Lerner, J. (1996). *The Government as Venture Capitalist: The Long-Run Impact of the SBIR Program*. NBER Working Paper 5753.

[42] Levin, R. C., Klevorick, A. K., Nelson, R. R., and Winter, S. G. (1984). *Survey Research on R&D Appropriability and Technological Opportunity: Part 1*. Working Paper, Yale University.

[43] Levin, R. C., Klevorick, A. K., Nelson, R. R., and Winter, S. G. (1987). Appropriating the returns from industrial research and development. *Brook. Pap. Econ. Act.* 1987, 783–820.

[44] Lichtenberg, F. R. (1987). The effect of government funding on private industrial research and development: a re-assessment. *J. Ind. Econ.* 36, 97–104.

[45] Link, A. N. (1987). "Technological change and productivity growth, in the series," in *Fundamentals of Pure and Applied Economics,* Vol. 13, ed. J. Lesourne (Paris: Harwood Academic Publishers).

[46] Link, A. N. (1997). *Public/Private Partnerships as a Tool to Support Industrial R&D: Experiences in the United States*. Report prepared for the Working Group on Innovation and Technology Policy. Paris: OECD.

[47] Mansfield, E. (1985). How rapidly does new industrial technology leak out? *J. Ind. Econ.* 34, 217–223.

[48] Matutes, C., Regibeau, P., and Rockett, K. (1996). Optimal patent design and the diffusion of innovations. *Rand J. Econ.* 27, 60–83.

[49] Mowery, D. (1995). "The practice of technology policy," in *Handbook of the Economics of Innovation and Technological Change*, ed. S. Paul (Oxford: Blackwell Publishers Ltd.).

[50] Mowery, D. C., and Rosenberg, N. (1993). "The U.S. national innovation system," in *National Innovation Systemsm*, ed. R. R. Nelson (Oxford: Oxford University Press).

[51] Nelson, R. (1961). Uncertainty, learning, and the economics of parallel research and development efforts. *Rev. Econ. Stat.* 43, 351–64.

[52] Reinganum, J. F. (1983). Uncertain innovation and the persistence of monopoly. *Am. Econ. Rev.* 73, 741–748.

[53] Reinganum, J. F. (1984). Preemptive patenting and the persistence of monopoly: reply. *Am. Econ. Rev.* 74, 251–253.

[54] Resnick, D. (1999). "Politics on the internet: the normalization of cyberspace," in *The Politics of Cyberspace*, eds C. Toulouse and T. W. Luke (London: Routledge).

[55] Scott, J. T. (1980). "Chapter 13-Corporate finance and market structure," in *Competition in the Open Economy: A Model Applied to Canada*, eds R. E. Caves, et al. (Cambridge: Harvard University Press), 325–359.

[56] Schumpeter, J. A. (1994). Capitalism, Socialism and Democracy. *London: Routledge.*

[57] Steinmueller, W. E. (1987). *International Joint Ventures in the Integrated Circuit Industry*. CEPR Publication No. 104.

[58] Stern, J. (1999). *The Ultimate Terrorists*. Cambridge, MA: Harvard University Press.

[59] United Nations (2010). Report of the Group of Governmental Experts on Developments in the Field of Information and Telecommunications in the Context of International Security, A/65/201, Study Series 33, 2010.

[60] Vonortas, N. S. (1994). Inter-firm cooperation with imperfectly appropriable research. *Int. J. Ind. Organ.* 12, 413–435.

[61] Weimann, G. (2004). *How Modern Terrorism Uses the Internet*. Washington, DC: United States Institute of Peace.

Index

C

Collaboration 7, 48, 66, 79

F

Faith 2, 16, 23, 34

I

Information and communications
    technology (ICT) 36, 60, 107
Information technology (IT) 37,
    56, 57

M

Mobile apps 14, 72, 73, 75

P

Policy 54, 71, 84, 87

S

Social networking 11, 12, 69

T

Technology 1, 7, 21,
    23, 32

W

Web 2.0 15, 31, 45, 56

About the Author

Ayse Kok received her MSc and Doctorate degree in Technology & Learning in University of Oxford in 2006. She participated in various research projects for UN, Nato and the EU and held various positions in global companies such as E&Y and BNPP. She worked as an adjunct faculty member at Bogazici University in her home town Istanbul, Turkey and participated as an invited speaker at various international conferences and published more than 50 articles. Ayse has also a second Masters degree in Tech Policy from Cambridge University. Currently, Ayse lives in Silicon Valley where she works as a contractor researcher for Google on user interaction design.